Lecture Notes in Mechanical Engineering

Lecture Notes in Mechanical Engineering (LNME) publishes the latest developments in Mechanical Engineering—quickly, informally and with high quality. Original research reported in proceedings and post-proceedings represents the core of LNME. Volumes published in LNME embrace all aspects, subfields and new challenges of mechanical engineering. Topics in the series include:

- Engineering Design
- Machinery and Machine Elements
- Mechanical Structures and Stress Analysis
- Automotive Engineering
- Engine Technology
- Aerospace Technology and Astronautics
- Nanotechnology and Microengineering
- Control, Robotics, Mechatronics
- MEMS
- Theoretical and Applied Mechanics
- Dynamical Systems, Control
- Fluid Mechanics
- Engineering Thermodynamics, Heat and Mass Transfer
- Manufacturing
- Precision Engineering, Instrumentation, Measurement
- Materials Engineering
- Tribology and Surface Technology

To submit a proposal or request further information, please contact the Springer Editor of your location:

China: Ms. Ella Zhang at ella.zhang@springer.com
India: Priya Vyas at priya.vyas@springer.com
Rest of Asia, Australia, New Zealand: Swati Meherishi at swati.meherishi@springer.com
All other countries: Dr. Leontina Di Cecco at Leontina.dicecco@springer.com

To submit a proposal for a monograph, please check our Springer Tracts in Mechanical Engineering at http://www.springer.com/series/11693 or contact Leontina.dicecco@springer.com

Indexed by SCOPUS. All books published in the series are submitted for consideration in Web of Science.

More information about this series at http://www.springer.com/series/11236

Ramesh K. Agarwal
Editor

Recent Advances in Manufacturing Engineering and Processes

Proceedings of ICMEP 2021

Springer

Editor
Ramesh K. Agarwal
Department of Mechanical Engineering
and Materials Science
Washington University in St. Louis
St. Louis, MO, USA

ISSN 2195-4356 ISSN 2195-4364 (electronic)
Lecture Notes in Mechanical Engineering
ISBN 978-981-16-3933-3 ISBN 978-981-16-3934-0 (eBook)
https://doi.org/10.1007/978-981-16-3934-0

This Springer imprint is published by the registered company Springer Nature Singapore Pte Ltd.
The registered company address is: 152 Beach Road, #21-01/04 Gateway East, Singapore 189721, Singapore

Organization

Conference Chairs

Tan Kiang Hwee, National University of Singapore, Singapore
C. W. Lim, City University of Hong Kong, Hong Kong
Ramesh K. Agarwal, Washington University in St. Louis, USA

Program Chairs

Kyoung Sun Moon, Yale University, USA
Ippei Maruyama, Nagoya University and The University of Tokyo, Japan

Publicity Committee

Galina Gorbacheva, Bauman Moscow State Technical University, Russia

Technical Committees

Mehmet Karakose, Firat University, Turkey
Silviu-Mihai Petrisor, Land Forces Academy from Sibiu, Romania
Constantin Dulucheanu, University "Stefan cel Mare" of Suceava, Romania
Zorica Veljkovic, University of Belgrade, Serbia
Radu Godina, NOVA University Lisbon, Portugal
Samad Nadimi Bavil Oliaei, Çankaya University, Turkey
Baris Burak Kanbur, Nanyang Technological University, Singapore

Panarat Rattanaphanee, Suranaree University of Technology, Thailand
Markus Brillinger, Graz University of Technology, Austria
Franco Concli, Free University of Bolzano, Italy
Charnnarong Saikaew, Khon Kaen University, Thailand
K. Abou-el-Hossein, Nelson Mandela Metropolitan University, South Africa
Quy Thu Le, National Research Institute of Mechanical Engineering, Vietnam

Secretariat

Jane Li, China

Preface

The 10th International Conference on Manufacturing Engineering and Processes 2021 (ICMEP 2021) was held virtually during March 11–14, 2021. Thank you for all the individuals who have supported and helped in organizing 10th ICMEP.

ICMEP is an annual conference organized with the intend of being a platform for researchers, engineers, academicians as well as industrial professionals from all over the world to present their research results and development activities in manufacturing engineering and process. This conference provides opportunities for the delegates to exchange new ideas and application experiences face to face, to establish business or research relations and to find global partners for future collaboration.

Due to COVID-19, the conference has to be held online instead of physically held in Singapore. Although we could not get together in Singapore, I believe it was still a pleasant experience and memory for all.

The conference was composed of 4 keynote and invited speeches and 12 parallel sessions. Participants are from France, Italy, Egypt, Thailand, turkey, Japan, China, and Poland, etc. Presenters actively participated in all sessions to make oral presentations and answer questions. All of them had done a great job.

The conference proceedings consist of 23 selected papers presented at ICMEP 2021, which include diverse topics covering recent trends in manufacturing engineering and processes. We hope that the conference proceedings will offer the readers a good insight into the future direction of research in this field.

We express our gratitude to our speakers, conference committees, authors, and listeners. We sincerely appreciate our conference committees spare their precious time to conduct the review process in order to maintain the quality of proceedings. Thanks to all participants' generous support and kind understanding! Hope all of you have had a good harvest during the conference, and we sincerely hope to meet you next year!

St. Louis, USA Ramesh K. Agarwal

Contents

Materials Science and Chemical Engineering

About the Editor

Prof. Ramesh K. Agarwal is William Palm Professor of Engineering, at Washington University in St. Louis, USA. Over a period of 45 years, Prof. Agarwal has worked in computational fluid dynamics (CFD), computational material science and manufacturing, and multi-disciplinary design and optimization. More recently, he has devoted some of his efforts in nanocomposites, shape memory alloys, and metamaterials. He is the author and co-author of over 600 publications and serves on the editorial board of more than 20 journals. Prof. Agarwal is a fellow of 25 societies including AIAA, IEEE, ASME, APS, IOM3, IET, RSC, etc.

About the Editor

Additive Manufacturing and Materials Forming

Hybridization of Fused Filament Fabrication Components by Stereolithographic Manufactured Thermoset Inserts

M. Baranowski, T. Schlotthauer, M. Netzer, P. Gönnheimer, S. Coutandin, J. Fleischer, and P. Middendorf

1 Introduction

Due to the development of new emission-free, electric powertrain concepts in the automotive industry, a change is taking place toward updated and upgradeable mobility concepts. The vision is an end-to-end digital production line for fast product development, but this also requires an adaptation of the associated production processes in order to enable a flexible but still economical fabrication. In consequence of the continuously rising speed of parts development, it is necessary to enable a fast and short-term adaptation of production lines with short setup times [1]. However, the degree of complexity of novel vehicle components is also increasing due to the requirements for installation space, weight, and additional functionalities. This leads to the need for new concepts and processes for future production technologies.

Additive manufacturing (AM) offers great advantages in terms of geometrical design freedom and fast development cycles and hence seen as an innovation driver for flexible production of future mobility concepts [2–4]. The basic principle usually consists of a layered build-up of the components [4]. Especially, fused filament fabrication (FFF) systems have become widely used in recent years due to its large and diverse range of thermoplastic materials and its rather easy handling. The application in an industrial context, however, is limited and mostly used for pro-

M. Baranowski (✉) · M. Netzer · P. Gönnheimer · S. Coutandin · J. Fleischer
Wbk Institute of Production Science, Karlsruhe Institute of Technology (KIT), Kaiserstraße 12, 76131 Karlsruhe, Germany
e-mail: michael.baranowski@kit.edu

T. Schlotthauer · P. Middendorf
Institute of Aircraft Design, University of Stuttgart, Pfaffenwaldring 31, 70569 Stuttgart, Germany

T. Schlotthauer
ARENA2036 Research Campus, Pfaffenwaldring 19, 70569 Stuttgart, Germany

R. K. Agarwal (ed.), *Recent Advances in Manufacturing Engineering and Processes*, Lecture Notes in Mechanical Engineering,
https://doi.org/10.1007/978-981-16-3934-0_1

totypes and production aids [5]. On the one hand, this is due to the anisotropic material behavior and the resulting typical structural failure between the build-up layers under load. On the other hand, caused by a low resolution and porous structure, there is usually insufficient accuracy to implement functional interfaces, such as threads, fits, sensors, or fluid-carrying areas. Complementary to this, stereolithographic (SL) manufacturing processes offer the possibility of producing highly precise and sealed structures, which however, due to their limited mechanical properties, are again mostly only used in the field of prototyping [6].

Therefore, this study will investigate the hybridization of both processes to exploit synergy effects and is compared with conventional metallic inserts as reference. In this context, hybridization means the intrinsic integration of subcomponents (metal/plastic parts) during the printing process. The aim of this study is to investigate the adherence properties of integrated metal and SL manufactured subcomponents.

2 Basics and State of Research and Technology

In additive manufacturing, beside selective laser sintering (SLS), two plastic processes have become established.

2.1 Fused Filament Fabrication (FFF) and Stereolithography (SL)

As in Fig. 1, FFF is a manufacturing process in which a workpiece (3) is built up layer by layer from a fusible plastic filament (2). A thermoplastic material is transported to a heated nozzle (1) and is melted there. The melted polymer filament is applied to a heated building platform (5, 6). Depending on the component complexity, support structures are used to create overhangs (4).

The main advantages and disadvantages of the FFF process are listed in the following Table 1 [6, 8].

According to Fig. 2, with SL objects (5) on a building platform (8, 9) are created from a liquid photopolymer resin (6) by curing this resin layer by layer with a laser or LED (2). Fresh resin is distributed on the building platform in layers with the help of a recoater (1). The exposure beam is directed to the target area by means of movable mirrors, so-called galvanometers (4) or digital micromirror devices (DMD). Exposure can be performed from above (top-down), as shown in Fig. 2, or from below (Bottom-up). Depending on the irradiation intensity and dose, the resin polymerizes in the exposed areas (3). Also here, depending on the component complexity, support structures are used to create overhangs (7).

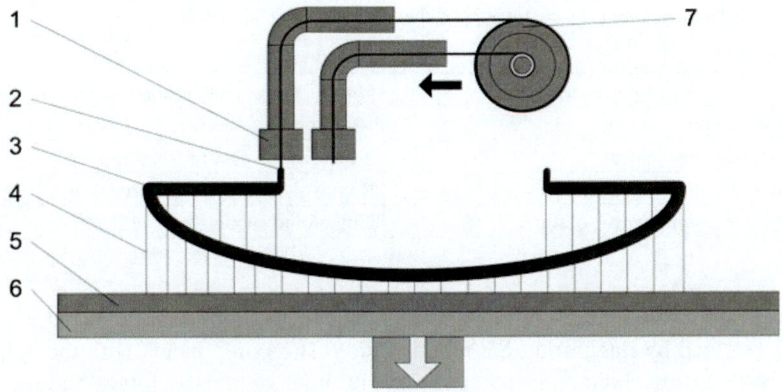

Fig. 1 Fused filament fabrication (FFF) process and its components [7]

Table 1 Advantages and disadvantages of the FFF process

Advantages	Disadvantages
Cheap	Ribbed surface
Mechanically resilient parts	Medium accuracy of production
Wide variety of materials	Porous structures

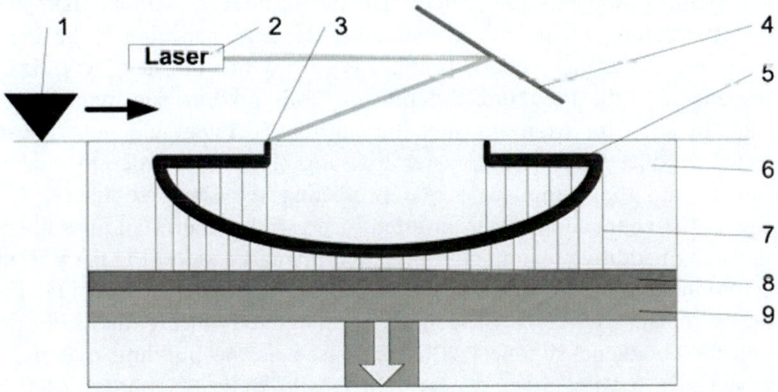

Fig. 2 Top-down stereolithographic (SL) process and its components [7]

Characteristic advantages and disadvantages of the process are listed in Table 2 [6, 9].

Component hybridization of both manufacturing processes can produce low-cost components (FFF) with locally adapted properties (SL). For example, the basic substrate of a plastic housing of a small electric motor can be manufactured at minimum cost using the FFF process. Parts with increased requirements for low

Table 2 Advantages and disadvantages of the SL process

Advantages	Disadvantages
High accuracy of production	Only UV hardenable polymers
Complex shaping	High production costs
Mechanically resilient parts	Necessity of post-processing step
Liquid and airtight	Removal of support material
Transparent components possible	Unicolored models

surface roughness, liquid, and airtightness as well as high manufacturing accuracy can be realized by integrating SL-manufactured sub-components with the help of a state-of-the-art machine concept. However, in order to transfer forces and moments into the housing, the interface, especially the adhesiveness, of both component partners must be investigated.

2.2 *Intrinsic Hybridization Concepts for Metallic Inserts*

The state of the art in research and technology already provides approaches to embed metallic inserts in component structures. The aim is to create a functional extension. For example, by integrating a load introduction element into the component structure, moments and forces can be introduced without damaging the component structure. Such load introduction elements can already be integrated during the manufacturing process. In this case, we define intrinsic hybridization of the component. In the literature, different methods for intrinsic hybridization are discussed. In [10], for example, metallic inserts are integrated into fiber-plastic composites produced by resin transfer molding (RTM) in order to enable load introduction into the composite while protecting the sensitive fibers. A major challenge is the connection of the inserted insert to the matrix and the fibers.

In the area of additive manufacturing, a hybridization concept in the FFF process for the production of metal/plastic composites was presented in [11]. For this purpose, a metal grid layer of copper is integrated during the FFF process. Acrylonitrile–butadiene–styrene (ABS) was used as the building material in the FFF process. In tensile tests, a maximum tensile force of approx. 2400 N was achieved depending on the lattice density of the metal.

In [12], the insertion of metallic inserts, especially nuts, into the FFF process was investigated. Besides the manufacturability and the geometric accuracy between metallic objects and FFF structures, adhesion plays a decisive role for the successful integration of an insert. Knoop et al. [12] investigated the adhesion with the help of the test specimen below (Fig. 3). The base substrate was printed up to the layer of integration, and then the part was inserted. After insertion, the printing process was continued, i.e., the insert was overprinted with two layers of polymer. In the course of this investigation, the screen angle, ribbon width, and spacing were varied.

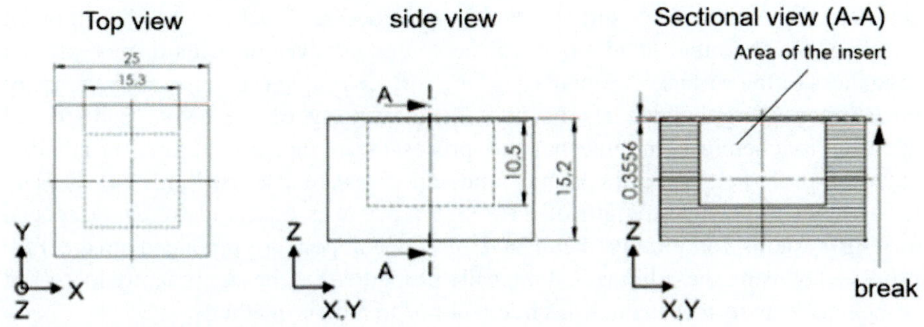

Fig. 3 Test specimen for testing adhesion [12]

Adhesion was evaluated on the basis of microscope images. A homogeneous inner grid without distortion leads to a high adhesion tendency when overprinting (Fig. 4 left). A high degree of adhesion was achieved by increasing the width of the strands. At imperfections, overprinting caused the molten filament to break off.

In [12], a strand width of ≥ 0.60 mm was specified. A larger strand width allows the nozzle to lay down a more robust strand and to deposit it more defined on the metal surface. The strand distance of 0 mm should be avoided. If the strand distance is negative, the deposited strands may overlap and thus be torn off (Fig. 4 right). A strongly positive strand distance leads to a lower connection of the deposited strand on the metal surface. The screen angle has no influence on the adhesion. Another important factor influencing the adhesion tendency of the plastic is the ratio between the surrounding FFF surface and the area of the metal insert. Small surfaces can be compensated by the over stretchability of the polymer strand. For a high adhesion quality, the surrounding FFF surface should be at least 50% larger than the insert surface to be overprinted.

Numerous methods for increasing the adhesion properties between plastic and metal components are presented in the literature. On the one hand, adhesion can be

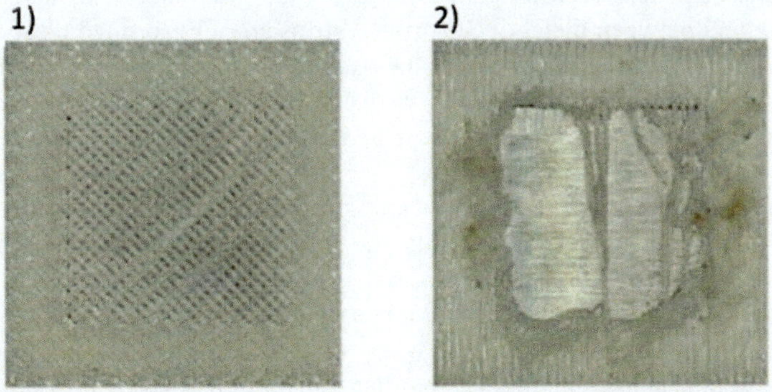

Fig. 4 Different grades of adhesion [12]

achieved by mechanical surface treatment processes such as sandblasting or grinding. On the other hand, chemical processes use lyes or acids to increase the roughness of the surface. An increase of the surface roughness causes an increase of the surface energy which results in a better wetting of the active partners. In addition to chemical and mechanical processes to increase surface roughness, adhesion promoters, special coatings, and adhesives are also used [13, 14]. In [14], an increase in tensile strength of 21.9 ± 1.1 MP was observed by using special coating systems for joining metal and plastic components produced in the FFF process. By using these binders, the tensile properties can be significantly increased compared to pure mechanical or chemical pre-treatment methods.

2.3 Research Deficit

The current state of research and technology includes some general approaches to significantly increase the adhesion properties between metal and plastic components. By using adhesion promoters, adhesives, special pre-treatment methods or coatings, the surface energy can be increased and thus an improvement in adhesion can be achieved. However, component hybridization between FFF and SL has not yet been investigated in the current state of research and technology. For this reason, the mechanical properties of thermoset acrylate SL inserts in FFF-manufactured thermoplastic polylactide (PLA) carrier components will be investigated in this study. A comparison with inserted metal strips in identical carrier components will be performed.

3 Experimental Setup and Results

3.1 Experimental Setup

The tensile shear test according to DIN EN 1465 was chosen to investigate the adhesive effect between metal or SL-printed strips and FFF-printed plastic samples. This standard is mainly used to assess the usability and quality of adhesive systems. Since different types of failure occur between the samples, the maximum tensile force is used to compare the results. The sample geometries used are shown in Fig. 5.

The metal and SL strips were pre-treated mechanically and chemically according to Table 3 in order to investigate only the influence of the surface condition. In addition, the strip surfaces were measured with a mobile tactile measuring device (ATP HRT-W5) to determine the influence of the surface roughness on the adhesion effect. A form fit via pin was selected for the comparison of the substance-to-substance bondings (Pin diameter 7 mm). The FFF samples were manufactured using an Ultimaker 2+. The following settings were selected: nozzle

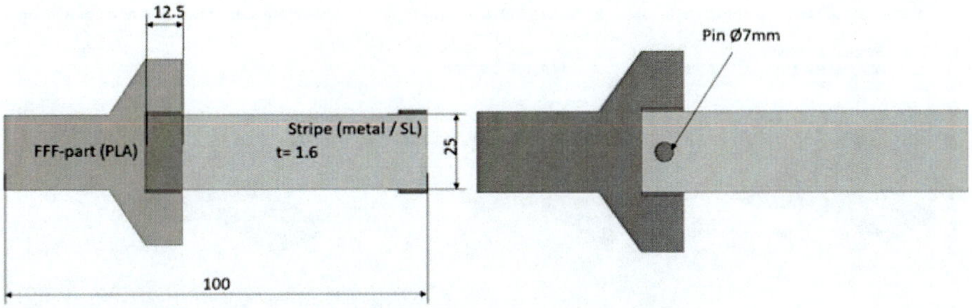

Fig. 5 Test samples with adhesive (left)/form fit (right) connection

Table 3 Pre-treatment of the metal/SL samples

Type of joining connection	Metal inserts	SL insert
Untreated	X	x
Refined	X	–
Pickled	X	–
Uncured	–	x
Sanded	X	x
Sand blasted	X	x
Form fit	X	x

diameter: 0.8 mm; layer thickness: 0.2 mm; printing temperature: 220 °C; filling degree: 40%; printing speed 40 mm/s; material: PLA (Ultimaker PLA); slicer: Cura. The process sequence for inserting the strips is shown in the following Fig. 6.

First the basic substrate of PLA is produced in the FFF process. Once a previously defined layer is reached, the printing process pauses. During the pause, the prepared metal/SL strips are inserted into the cavity. After the strips have been inserted, the building job is continued, and the printer overprints the strip surface (25 mm × 12.5 mm = 312.5 mm^2).

The structural steel strips are folded parts that have been cut out of a metal sheet according to DIN EN 1465. The SL inserts are manufactured out of the thermoset acrylate photopolymer Type D Standard from Druckwege GmbH (Hennef, Germany) with a D30II printer from Rapid Shape GmbH (Heimsheim, Germany) (Fig. 7 left). In order to investigate the influence of the residual reactivity and the resulting tacky surface of the photopolymer, a set of the samples was inserted into the FFF structure without post-exposure after the SL printing process (referred as "uncured"). This set has been post-exposed afterward together with the FFF structure (Fig. 7 right). All other samples are exposed from each side with a dose of 5 J/cm inside the Opsytec BS-02 UV chamber (Opsytec Dr. Gröbel GmbH, Ettlingen, Germany) before inserted to the FFF component. The camber was equipped with 8 UVA mercury vapor lamps and was controlled via calibrated RM 12 radiometer sensors to a total radiation intensity of 6.7 mW/cm^2. This results into a tensile strength of 36.6 MPa, Young's modulus of 1.28 GPa, and 4.5% elongation at break according to the manufactures datasheet [15]. In order to investigate the

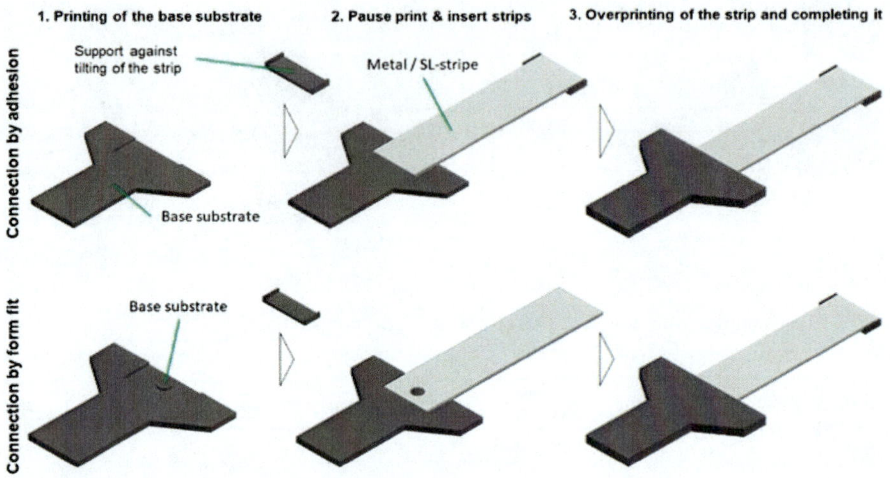

Fig. 6 Process sequence for inserting the strips (metal/SL) into the base substrate (PLA)

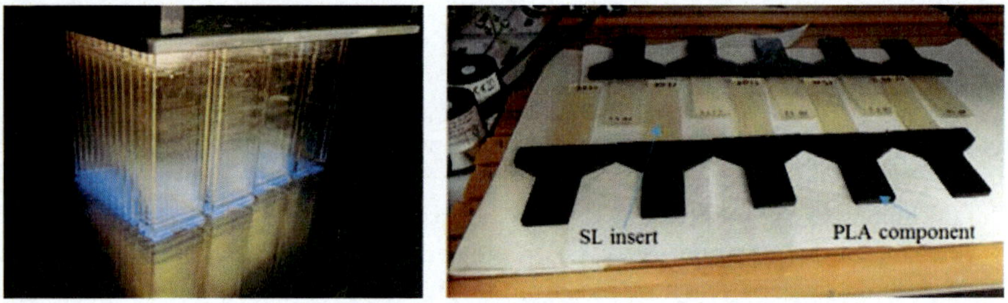

Fig. 7 SL insert sample production. Left: UV-light based production of the initial samples. Right: Post-processing in UV camber to reach complete curing.

influence of surface pre-treatment, one set each was treated by means of sand-blasting and sandpaper (grain size 200). A form-fit connection was examined by adding a hole with 7 mm diameter, which was already included during the printing process (Fig. 6).

The tensile tests were performed using a tensile testing machine from Zwick/Roel (Load Cell Xforce K 20kN). A test speed of 2 mm/min and a preload of 1 N was selected.

3.2 Results

The surface roughness values were determined for the pre-treatment methods listed in Table 3. Figure 8 shows the averaged roughness (R_z) of the metal and SL samples. For metals, the highest roughness was produced by pickling. A treatment

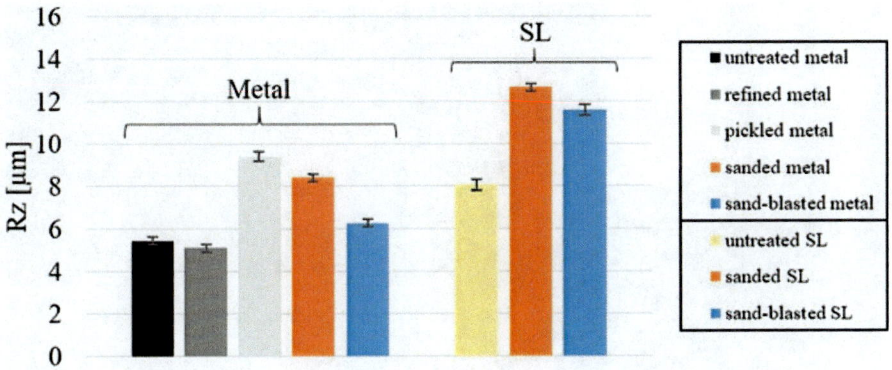

Fig. 8 Comparison of roughness values of metal an SL insert samples

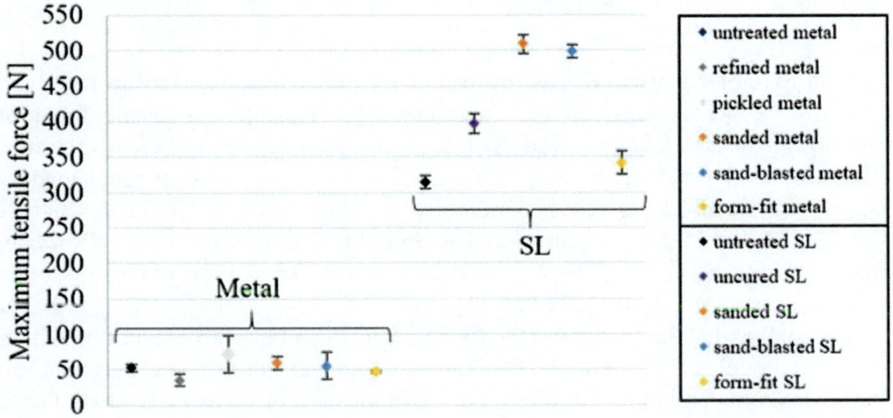

Fig. 9 Results of lap shear tests with different pre-treated metal (left) and SL (right) inserts in combination with PLA during FFF process

by cleaning causes a minimal reduction of the roughness, because this process step leads to a smoothing of the surface. If the roughness values of metal and SL are compared, the values of the SL samples are at a higher level due to the softer material after sanding and sandblasting.

With increasing roughness of the metal samples, a small increase of the maximum tensile force can be observed (Fig. 9). However, the differences between the untreated metal insert and the four surface treatments (refined, pickled, sanded, sand blasted) show no statistically significant difference because of the high variance, except for the refined sample. The maximum tensile forces are between 35.45 ± 8.37 and 72.00 ± 26.13 N. The metal form-fit shows a much smaller variance of the results with 47.20 ± 1.95 N, but only a higher maximum tensile strength than the refined sample. All metal samples failed due to a pull out of the insert, which indicates a bad adhesion to the PLA.

Fig. 10 Lap shear specimens after tests. **a** and **b**: PLA/SL specimen with fracture plane in SL component at transition between both materials. **c**: Traces of FFF nozzle on SL surface

If an SL-printed insert is used instead, a significantly higher tensile force can generally be achieved. Even for the untreated SL sample, a tensile force of 314.94 ± 9.91 N can be observed. Also a greater influence of the increased surface roughness can be seen compared to the metal inserts. If the surface is roughened by sanding and sandblasting, the maximum tensile force is up to 509.12 ± 13.45 N and 498.52 ± 8.99 N, respectively. The initially uncured specimen allows an increase to 396.88 ± 14.32 N, but, like the form-fit variant, is significantly lower than the sanded and sand-blasted specimens.

It is notable that the fracture of the FFF/SL specimen has always occurred within the SL component. For the untreated, uncured, and form-fit samples, this occurred mainly in the transition region (Fig. 10a, b). In the case of sanded and sand-blasted specimens, a fracture usually occurs within the clamping jaws. By means of a pulled out SL specimen, the traces of the FFF nozzle paths can be identified on the surface (Fig. 10c). This shows that there are micromechanical changes in the surface structure, which improves the adhesion.

By considering the cross-sectional area of the SL insert, the sanded sample achieves a tensile strength of approximately 12.73 ± 0.34 MPa. This is approx. 1/3 of the tensile strength of the purely printed polymer [15]. It is noticeable that the elongation at break was only 0.73 ± 0.11%. According to the manufacturers specification, the elongation at break should be 4.5% [15]. This indicates that the interface would have been capable of withstanding a greater tensile load. This can be explained by early SL material failure due to stress concentrations at the transition between the two materials and the superposition of compression and tensile load with early crack formation in the clamping jaws.

4 Conclusion and Outlook

The conducted study shows the feasibility of a FFF structure hybridization by integration of SL printed inserts. This enables the synergy of material properties, e.g., fluid-carrying cooling channels made of tight and resistant SL materials in combination with low-cost FFF housings, and hence allows an extension of the application spectrum of additively manufactured plastic components.

For optimum adhesion, the investigations have shown that the surface of the thermoset acrylate SL material should be mechanical treated with sandpaper or sandblasting to produce a significant increase in roughness. In direct comparison to metallic inserts, SL inserts generally offer better adhesion to thermoplastic PLA, whereby the mechanical surface treatment has a much greater influence than the form-fit. This allowed up to 1/3 of the maximum tensile strength of the SL material to be applied before premature failure in the clamping jaws occurred. Future studies should therefore perform material testing using additional force introduction elements on the specimen or a more ductile SL material to identify the maximum transmittable forces.

Another further research aspect is the fully automatic insertion of the SL components to achieve an increase in production efficiency. In this context, special challenges of control engineering as well as the accuracy of robot and kinematics have to be solved. Especially when combining complex shapes, such as multi-bent cooling channels, high accuracy during placement is required. This should be tested in the next steps on component level, considering the influence of warpage on insert integration and the usage of subtractive surface pre-treatments.

Acknowledgements The authors would like to thank the Ministry of Science, Research and Arts of the Federal State of Baden-Württemberg, Germany for the financial support of the projects within the InnovationsCampus Mobilität der Zukunft and ARENA2036 for the possibility to conduct the investigations within the research campus.

References

1. Abele E, Reinhart G (2011) Future of production: Challenges, fields of research, opportunities. Carl Hanser Fachbuchverlag, s.l.
2. Baumann F, Scholz J, Fleischer J (2017) Investigation of a new approach for additively manufactured continuous fiber-reinforced polymers 66:323
3. Spiller Q, Fleischer J (2018) Additive manufacturing of metal components with the ARBURG plastic freeforming process 67:225
4. Richard HA, Schramm B, Zipsner T (eds) (2017) Additive manufacturing of components and structures. Springer Vieweg
5. Klahn C, Meboldt M, Fontana F, Leutenecker-Twelsiek B, Jansen J (eds) (2018) Development and design for additive production: basics and methods for the use in industrial end customer products, 1st edn. Vogel Business Media
6. Gebhardt A (2016) Additive manufacturing processes: additive manufacturing and 3D printing for prototyping—Tooling—Production, 5th edn. Hanser, München

7. VDI 3405 (2014) Additive manufacturing processes: basics, definitions, processes, Berlin. Beuth Verlag GmbH 25.020 (3405). Accessed 10 Nov 2020
8. Fuhrmann M (2018) Quality-oriented model-based process parameter optimization for fused deposition modeling
9. Zhou JG, Herscovici D, Chen CC (2000) Parametric process optimization to improve the accuracy of rapid prototyped stereolithography parts 40:363
10. Gebhardt J, Schwennen J, Lorenz F, Fleischer J (2020) Structure optimisation of metallic load introduction elements embedded in CFRP. https://link.springer.com/article/https://doi.org/10.1007/s11740-018-0820-5. Accessed 5 Nov 2020.774Z
11. Butt J, Shirvani H (2018) Experimental analysis of metal/plastic composites made by a new hybrid method 22:216
12. Knoop F, Köhler M, Schöppner V, Lieneke T et al (2018) Development of design rules for hybrid components: integration of metallic inserts in FDM structures
13. Kweon J-H, Jung J-W, Kim T-H, Choi J-H et al (2006) Failure of carbon composite-to-aluminum joints with combined mechanical fastening and adhesive bonding 75:192
14. Falck R, Goushegir SM, dos Santos JF, Amancio-Filho ST (2018) AddJoining: a novel additive manufacturing approach for layered metal-polymer hybrid structures 217:211
15. DruckWege. DLP Resin for 3D printers from Germany—DruckWege. https://druckwege.de/3d-druck-resin. Accessed 10 Nov 2020

Multiobjective Optimization of 3D-Printed Injection Molds via Hybrid Latin Hypercube Sampling-Delaunay Triangulation Approach

Baris Burak Kanbur, Suping Shen, Volkan Kumtepeli, Yi Zhou, and Fei Duan

1 Introduction

Conformal cooling channels (CCCs) are additive manufacturing-based channel structures implemented in the plastic injection mold geometries for manufacturing of complex plastic models [1]. Unlike the straight cooling channels manufactured by traditional machining tools, the CCCs can follow the complex pathways in the injection mold geometry thanks to the printing flexibilities of the additive manufacturing tools [2]. Therefore, they can achieve a better temperature uniformity that decreases the thermal gradients on the injection mold geometry from the first moment of the injection to the end of the ejection process. [3]. However, due to their complex designs, they also result in higher pressure drops and mechanical stress. Alternatively, conformal cooling cavities can provide an extremely shorter cooling time, but their mechanical performances are significantly weaker than the CCCs [4]. In an overall view, the design of CCCs has multiple objectives that are related to the structural, operating, and performance parameters so that single or multiobjective optimization studies are performed for the best trade-off point decision [5]. Our group also carried out multiobjective optimization studies considering the multiple objectives such as thermal gradients, cooling time, and pressure drop [6, 7]. Up to now, existing studies have performed single/multiobjective optimization with one or 2–3 design variables which allow them to create design spaces easily. Thus, tradi-

B. B. Kanbur (✉) · S. Shen · Y. Zhou
Singapore Centre for 3D Printing, NTU, Singapore 639798, Singapore
e-mail: bbkanbur@ntu.edu.sg

B. B. Kanbur · S. Shen · Y. Zhou · F. Duan
School of Mechanical and Aerospace Engineering, NTU, Singapore 639798, Singapore
e-mail: feiduan@ntu.edu.sg

V. Kumtepeli
Energy Research Institute, NTU, Singapore 637141, Singapore

R. K. Agarwal (ed.), *Recent Advances in Manufacturing Engineering and Processes*, Lecture Notes in Mechanical Engineering,
https://doi.org/10.1007/978-981-16-3934-0_2

tional full factorial design or similar methods have been efficiently applied to create design space; then, the model interpolation (a.k.a. meta-model) step has been performed with negligible computational effort (time and costs). However, it is a fact that the CCCs have complex designs with a higher number of design variables that make the traditional full factorial design method challenging for generating a meta-model. The reason is the high number of design variables requires enormous design space which will be costly and time-consuming. In the literature, the Latin Hypercube Sampling (LHC) method is proposed instead of the full factorial design since the LHC technique is a statistical sampling method that requires fewer samples for more design variables. To this end, this study presents a CCC-included injection mold geometry that has eight different design variables. The LHC technique is used for sample distribution in a multi-dimensional variable environment; then, the meta-model (interpolation) step is completed via Delaunay Triangulation (DT) method, which is more feasible in higher dimensions [8] and multiple objectives compared to currently performed the Kriging technique. The combined LHC-DT method has not been applied before as far as we know; thus, we aim to present this hybrid approach not only for the current design but also for other additive manufacturing-based processes, which include multi-dimensional design spaces. The results of the hybrid LHC-DT are sent to the multiobjective optimization tool in the MATLAB environment and then the Pareto frontier plots are generated for multi-criteria decision-making.

2 Materials and Methods

The proposed injection mold geometry is shown in Fig. 1 with eight different design variables: angle (α, between 0° and 90°), diameter (D, between 1 and 4 mm), height (H, between 5 and 20 mm), arc radius (r, between 8 and 50 mm), aspect ratio (a/b, between 0.3 and 3), and the distance values of x, y, and z (between 0.5 and 5 mm). These eight design variables define the CCC structure in the mold geometry and make the required design space extremely large that is costly and time-consuming via traditional full factorial design; hence, we applied the LHC technique, which requires a smaller number of sampling data thanks to its statistical background, in the MATLAB environment (*lhsdesign* command). According to the LHC, we defined 134 different simulation runs and then the simulations were performed via ANSYS CFD tool. Since the plastic injection is a transient process, we defined transient boundary conditions for the model with the time step of 0.1 s. The main design criteria were (i) the plastic injection temperature is 608 K and (ii) the cooling time is constant at 3.2 s. The mesh structure was the tetrahedral mesh with a size of 3 millions. The simulations were conducted to observe the objectives of (i) difference between the maximum and minimum temperature values at the internal wall, (ii) maximum temperature value at the plastic interface, and iii) pressure drop.

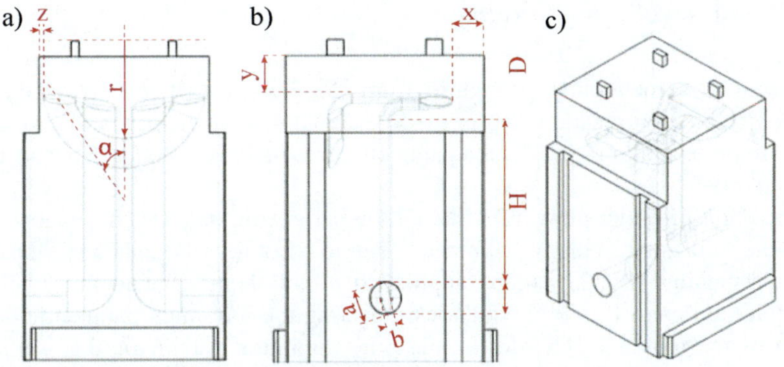

Fig. 1 3D model of the injection mold: **a** front view, **b** side view, and **c** isometric view

Once all simulations were completed, we created four different parameter sets with the simulations 1–30, 1–60, 1–90, and 1–134. The main reason behind the selection of four different parametric sets is observing and comparing the multi-objective optimization trends because each simulation simply means computational cost and time. Thus, we tried different parameter sets and compare their optimization results. If the optimization results are seen similar for the parametric sets; we, inherently, choose the smallest parametric set since it requires less computation cost and time. For each parametric set, we applied the Delaunay Triangulation (DT) technique in the MATLAB environment in order to accurately create the meta-model at higher dimensions because the current design variables result in a huge design space that cannot be efficiently interpolated via the Kriging approach. After completing the DT-based meta-model, we defined the meta-model as an initial population of the genetic algorithm for the above-given multiple objectives in the multiobjective optimization tool, which presents the Pareto frontier plot, in the MATLAB. By using the optimal points in the Pareto frontier plot, we can define the best trade-off points according to the defined objective weightage of the weighted-sum (WS) method. The multiobjective optimization problem and the weighted-sum approach for decision-making are presented in Eqs. 1 and 2, respectively.

$$\mathrm{MOF} = \min(f(\Delta T@\mathrm{Int.\,Wall}), f(\mathrm{Max.\,T}), f(\mathrm{Pres.\,Drop})) \tag{1}$$

$$\mathrm{WS} = a \cdot f^{*}(\Delta T@\mathrm{Int.\,Wall}) + b \cdot f^{*}(\mathrm{Max.}\,T) + c \cdot f^{*}(\mathrm{Pres.\,Drop}) \tag{2}$$

where $a, b, \text{and}\, c$ are the weights of the normalized functions (f^*) of the multiple objectives. The objective function normalization is performed with the utopia (f^U) and pseudo-nadir (f^{PN}) points of the Pareto frontier plot; $f^* = [f(x) - f^U(x)]/[f^{\mathrm{PN}}(x) - f^U(x)]$, where $f(x)$ is the function value obtained from the Pareto frontier data set in the multiobjective optimization procedure [7].

3 Results and Discussions

Following the mentioned procedure from the hybrid LHC-DT method to the multiobjective optimization step in the MATLAB environment, the Pareto frontier plots are projected in Fig. 2 for the parametric sets of 1–30, 1–60, 1–90, and 1–134, respectively.

The Pareto frontier plots infer that all the parametric data sets have close trends and they present clearly apparent Pareto frontier boundaries; thus, the decision-making (Eq. 2) can be applied to all of them. Due to the similarities between the Pareto frontier boundaries, selecting the minimum parametric data set (1–30) is the most feasible idea because the minimum parametric data set simply means less number of computational loads (both time and cost). The close Pareto frontier boundaries make the analysis of the best trade-off point very similar for the parametric data sets from 1–30 to 1–134. According to the parametric data set of 1–30, the best trade-off points are 40.68 K, 327.39 K, and 0.66 bar for ΔT@Int. Wall, Max. T, and the Pres. Drop, respectively, for the equal weightage for all objectives. Apart from the equal weightage scenario, we present the design variables for the best trade-off point of (i) with 80% weightage, (ii) with 80% weightage, and (iii) with 80% weightage as listed in Table 1.

Table 1 shows that the design variables for the best trade-off points of all scenarios are very close to one another so that they can be assumed the same values when we consider the sensitivity of the 3D printer. Among all variables, the biggest difference is observed for the angle (α), but it is, actually, also negligible as can be seen in Table 1.

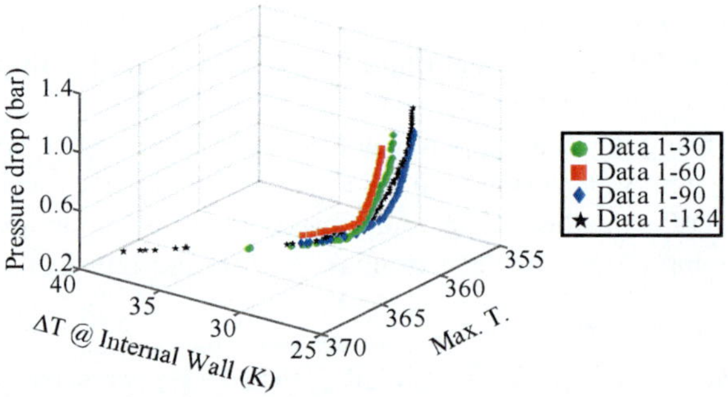

Fig. 2 Pareto frontier plots of different data set scenarios

Table 1 Design variables according to the best trade-off points of the data set of 1–30

Main objective	*A*	*D*	*H*	*R*	a/b	*X*	*y*	*z*
ΔT@Int. Wall	37.29	3.72	16.91	35.69	0.75	3.20	1.19	2.95
Max. T	35.78	3.68	16.14	35.48	0.77	3.18	1.02	3.04
Pres. drop	39.63	3.73	17.17	35.70	0.75	3.18	1.29	2.97
Equal weight	35.78	3.68	16.14	35.48	0.77	3.18	1.02	3.04

4 Conclusions

The present study developed a new hybrid LHC-DT approach for performing the multiobjective optimization of a complex 3D-printed injection mold geometry with three different objective functions. CFD simulations were carried out for 134 different runs and the results were used for meta-model creation to be evaluated in the multiobjective optimization. The results showed that a reliable Pareto frontier plot was obtained with the minimum parametric data set of 1–30 so that it can allow time and cost savings in the computational load compared to other data sets. The Pareto frontier boundaries were also close to one another. The multiobjective optimization study with main objective scenarios of equal weightage, Max. T, ΔT@Int. Wall, and Pres. Drop presented very similar design variables. The best trade-off point was observed at 28.88 K, 361.35 K, and 0.46 bar for the objectives of ΔT@Int. Wall, Max. T, and Pres. Drop, respectively.

Acknowledgements The authors would like to gratefully acknowledge the support of the Singapore Centre for 3D Printing (SC3DP) and School of Mechanical and Aerospace Engineering for this original research work. B.B. Kanbur is the Mistletoe Research Fellow granted by the Momental Foundation.

References

1. Kanbur BB, Shen S, Duan F (2020) Design and optimization of conformal cooling channels for injection molding: a review. Int J Adv Manuf Technol 106:3253–3271
2. Shinde MS, Ashtankar KM, Kuthe AM, Dahake SW, Mawale MB (2018) Direct rapid manufacturing of molds with conformal cooling channels. Rapid Prototyping J 24:1347–1364
3. Shen S, Kanbur BB, Zhou Y, Duan F (2020) Thermal and mechanical analysis for conformal cooling channel in plastic injection molding. Mater Today: Proc 28:396–401
4. Kanbur BB, Shen S, Zhou Y, Duan F (2020) Thermal and mechanical simulations of the lattice structures in the conformal cooling cavities for 3D printed injection molds. Mater Today: Proc 28:379–383
5. Kitayama S, Miyakawa H, Takano M, Aiba S (2017) Multi-objective optimization of injection molding process parameters for short cycle time and warpage reduction using conformal cooling channel. Int J Adv Manuf Technol 88:1735–1744
6. Kanbur BB, Shen S, Zhou Y, Duan F (2020) Neural network-integrated multiobjective optimization of the 3D-printed conformal cooling channels. In: 2020 5th international conference on smart and sustainable technologies (SpliTech), IEEE, Split, Crotia
7. Shen S, Kanbur BB, Zhou Y, Duan F (2020) Thermal and mechanical assessments of the 3D-printed conformal cooling channels: computational analysis and multi-objective optimization. J Mater Eng Perform 29:8261–8270
8. Kumtepeli V, Hesse HC, Schimpe M, Tripathi A, Wang Y, Jossen A (2020) Energy arbitrage optimization with battery storage: 3D-MILP for electro-thermal performance and semi-empirical aging models. IEEE Access 8:204325–204341

Mechanical Properties of Polycarbonate Urethane (PCU) for Orthopedic Devices: Fabrication with Compression Molding

W. D. Lestari, R. Ismail, J. Jamari, and A. P. Bayuseno

1 Introduction

The lifespan of UHMWPE joints is often limited to 15–20 years [1]. The limitation of UHMWPE is that the material is stiff enough that it cannot imitate the viscoelastic properties of real articular cartilage such as those found in the hip joint. Based on these problems, there are two basic approaches that can be used to extend the life of the artificial hip joint. The first option is to reduce the wear rate of the UHMWPE ace tabular liner. The second option is to use an alternative substitute for UHMWPE which is similar in nature to real cartilage. The second option is a solution taken by many researchers recently, by looking for alternative materials that have mechanical properties similar to original cartilage.

Recent studies found the polycarbonate urethane (PCU) to be the most promising as an alternative bearing [2–6] solution to UHMWPE in orthopedic load bearing applications in the artificial hip prosthesis [7]. PCU is an attractive material which has combination of elasticity with durability, wear and corrosion resistance, toughness, mechanical stability, good tribological properties, good compatibility, biologically stable polymer with a low-elastic modulus (10–100 Mpa) [8, 9] that is similar to natural cartilage [10, 11]. Concerning to get the required properties of PCU, there are a few studies on the fabricated of this thermoplastic material. Miller et al. [12] assessed three processing methods of PCU (injection molding, compression molded, and 3D printed) to acquire the best material for long life performance of PCU implants in load bearing applications. The overall objective of this study is to evaluate the effect of

W. D. Lestari (✉)
Department of Mechanical Engineering, Faculty of Engineering, University of Pembangunan Nasional "Veteran" Jawa Timur, Surabaya 60294, Indonesia
e-mail: wahyu.dwi.tm@upnjatim.ac.id

R. Ismail · J. Jamari · A. P. Bayuseno
Department of Mechanical Engineering, Faculty of Engineering, Diponegoro University, Semarang 50275, Indonesia

R. K. Agarwal (ed.), *Recent Advances in Manufacturing Engineering and Processes*, Lecture Notes in Mechanical Engineering,
https://doi.org/10.1007/978-981-16-3934-0_3

compression molding PCU on the mechanical properties and material characterization for orthopedic devices such as bearing material of artificial hip joint application.

2 Materials and Method

2.1 Materials and Processing

Compliant polycarbonate urethane (PCU) Bionate 80A in pellet form. Processing methods utilized in this study by compression molding. The compression molding parameters in this study are shown in Table 1, as for the process shown in Fig. 1.

2.2 Mechanical Properties

The mechanical properties that use in this study were tensile test, compression test, and hardness test. Tensile testing is carried out based on the standard DIN EN ISO 527-2 type-1BA, while the compression test refers to ASTM D 6147-97 method A and research from Beckmann [13]. The hardness measurement in this study was taken using a Vickers hardness tester (Mitutoyo HM 200).

2.3 Materials Characterization

The morphological analysis of PCU specimens from the compression molding process was carried out using scanning electron microscopy (SEM). Furthermore, the materials characterization that is carried out were absorption test and density. Water absorption testing is carried out based on ASTM D750 standards and research conducted by Geary [14]. In addition, the density test is carried out according to the ASTM D792 standard.

Table 1 The compression molding parameter of PCU with a holding time of 30 s

Speciments	Melt temperature (°C)	Pressure (Kgf)
Speciment 1/PCU A	180	65
Speciment 2/PCU B	190	60
Speciment 3/PCU C	200	55

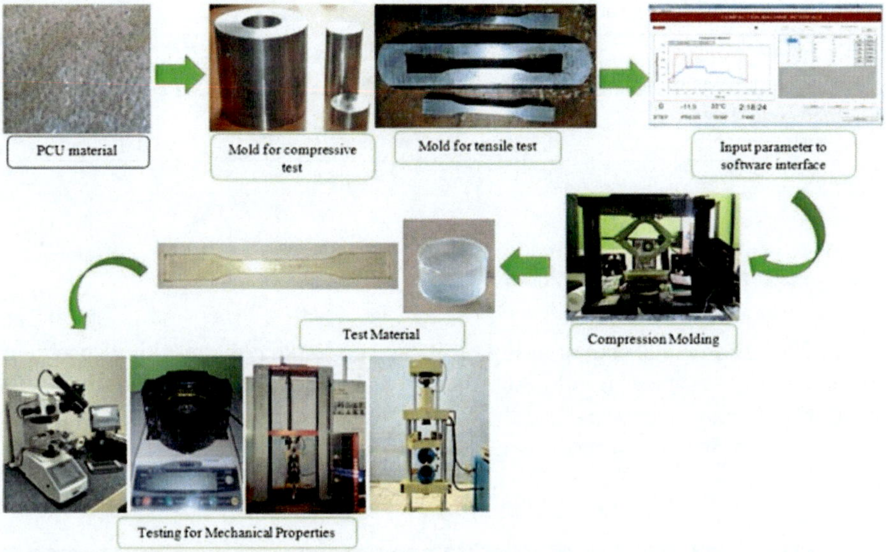

Fig. 1 The compression molding process of PCU material

3 Results and Discussion

3.1 Mechanical Properties

Tensile Test. The tensile test results show that the average modulus of elasticity for PCU A specimens is 11.36 MPa with a strain of 402.93%, PCU B specimens 10.86 MPa with a strain of 402.94%, and PCU specimen C 10.82 MPa with a strain of 402.93%.

Compressive Test. The results of the compressive test on each specimen in this study are shown in Table 2.

Hardness. Hardness testing is based on ASTM D2240 standards. The standard required for PCU is to have a hardness value of 83–85.5 shore A. The results of PCU material hardness testing with a compression molding process are shown in Table 3.

Table 2 Compressive test results

Name	Diameter	High	Mod. Elastic	Stress	Max. strain
Parameter			Strain 0.5–0.25%	30 min	Entire areas
Unit	mm	mm	MPa	MPa	%
PCU A	29.72	5.92	0.1760	1.7143	25.054
PCU B	29.84	5.92	0.1145	1.4077	25.055
PCU C	29.66	6.05	0.1168	1.0928	25.054

Table 3 Results of hardness testing of PCU specimens

PCU specimen	Average hardness (HV)	Hardness (Shore A)	Hardness Standards (Shore A)	Differentiation (%)
Model A	807.0	89.3	83–85.5	4.44
Model B	760.4	85.6	83–85.5	0.01
Model C	686.6	80.3	83–85.5	

3.2 Material Characterization

SEM. The SEM results shown in Fig. 2, it can be seen that the resulting specimens still have rough surface. In addition, there are also waves which are probably the result of the pressing process and the melting temperature is not suitable, so that it has not yet produced a homogeneous surface.

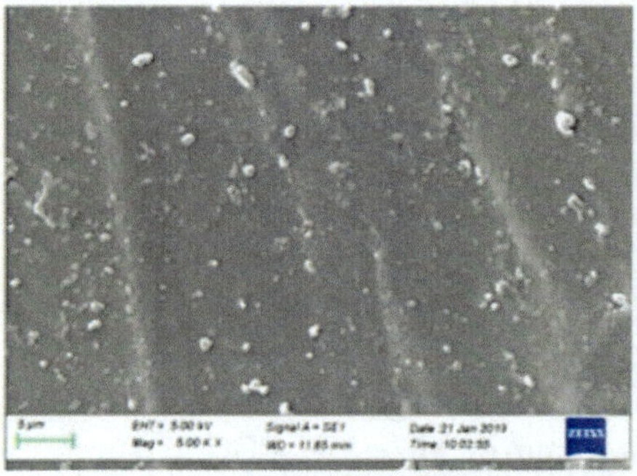

Polycarbonate Urethane 180°C (PCU A)

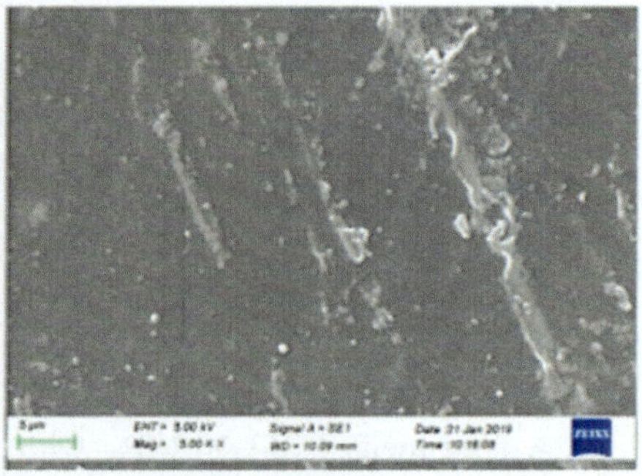

Polycarbonate Urethane 190°C (PCU B)

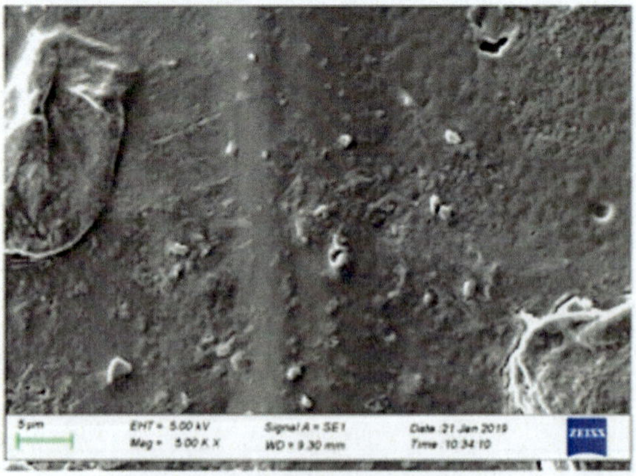

Polycarbonate Urethane 200°C (PCU C)

Fig. 2 Morphology of PCU specimen seen by SEM

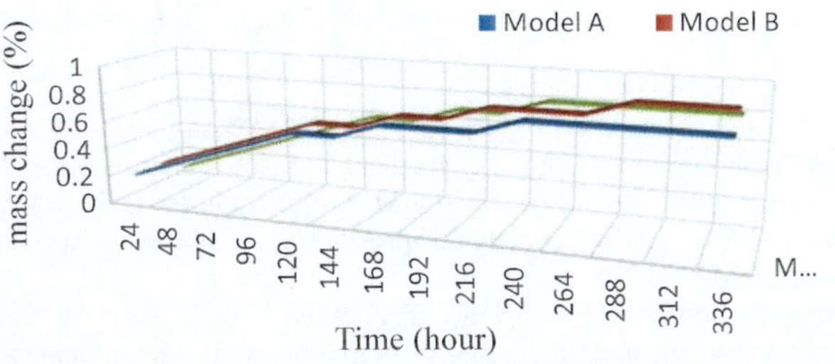

Fig. 3 The absorption test PCU material from compression molding

Tabel 4 Density test results of solid PCU specimens

PCU specimen	Density standard (g/cc)	Density obtained (g/cc)	Differentiation (%)
Model A	1.19	1.18	0.84
Model B	1.19	1.19	0.00
Model C	1.19	1.18	

Water Absorption. The standard required for PCU is having the ability to absorb water by 0.85–1.2%. Absorption test results are then made a graph of mass and time increment (see Fig. 3). Based on graph, it can be seen that the PCU A and PCU C specimens have the same total mass gain. As for PCU B specimens, the mass gain was smaller than PCU A and PCU B specimens.

Density. The standard required for a PCU is to have a density or density value of 1.19 g/cc. Based on Table 4, it can be seen that the standard density and density of the PCU specimens made by the compression molding process are relatively the same, with a difference of only 0.01 g/cc in model A and model C, whereas in model B the density is exactly the same, 1.19 g/cc.

4 Conclusion

Based on the experiment, it was concluded that the appropriate temperature for the PCU compression molding process was between 180 and 200 °C with a pressure of 65–55 kgf and a holding time of 0–0.5 min. The results of characterization in the form of density and water absorption tests were also in accordance with existing standards. As for the mechanical testing, it is still not maximal, but also in accordance with the existing standard range. Based on this, it can be said that the PCU material manufacturing method by means of compression molding can be adapted to process PCU material for medical equipment purposes.

References

1. Shahemi N, Liza S, Abbas AA, Merican AM (2018) Long-term wear failure analysis of uhmwpe acetabular cup in total hip replacement. J Mech Behav Biomed Mater 87:1–9
2. Carbone A, Howie DW, McGee M, Pearcy M, Smith N, Jones E (2006) Aging performance of a compliant layer bearing acetabular prosthesis in an ovine hip arthroplasty model. J Arthroplasty 21(6):899–906
3. Khan I, Smith N, Jones E, Finch DS, Elizabeth R (2005) Analysis and evaluation of a biomedical polycarbonate urethane tested in an in vitro study and an ovine arthroplasty model. Part I : Mater Sel Eval Biomater 26:621–631
4. Khan I, Smith N, Jones E, Finch DS, Elizabeth R (2005) Analysis and evaluation of a biomedical polycarbonate urethane tested in an in vitro study and an ovine arthroplasty model. Part II: In Vivo Invest Biomater 26:633–643
5. Scholes SC, Unsworth A, Blamey JM, Burgess IC, Jones E, Smith N (2005) Design aspects of compliant, soft layer bearings for an experimental hip prosthesis. Proc IMechE Part H J Eng Med 219:79–87
6. Scholes SC, Burgess IC, Marsden HR, Unsworth A, Jones E, Smith N (2006) Compliant layer acetabular cups: friction testing of a range of materials and designs for a new generation of prosthesis that mimics the natural joint. Proc IMechE Part H J Eng Med 220:583–596
7. Smith RA, Maghsoodpour A, Hallab NJ (2009) In vivo response to cross-linked polyethylene and polycarbonate-urethane particles. J Biomed Mater Res Part A, 227–234
8. Christiane A, Vrancken T, Buma P, Van Tienen TG (2013) Synthetic meniscus replacement: a review. Int Orthop 37:291–299
9. Sonntag R, Reinders J, Kretzer JP (2012) What's next? Alternative materials for articulation in total joint replacement. Acta Biomater 8(7):2434–2441
10. Baliga R, Shenoy S, Rao S, Bhat V (2013) Synthesis and mechanical characterization of C. N. F reinforced PCU composite. Int J Chem Environ Biol Sci 1(4):697–700
11. Pinchuk LS, Nikolaev VI, Tsvetkova EA (2006) Tribology and biophysics of artificial joints
12. Miller AT, Safranski DL, Smith KE, Sycks DG, Guldberg RE, Gall K (2017) Fatigue of injection molded and 3D printed polycarbonate urethane in solution. Polymer (Guildf) 108:121–134
13. Beckmann A, Heider Y, Sto M, Markert B (2018) Assessment of the viscoelastic mechanical properties of polycarbonate urethane for medical devices. J Mech Behav Biomed Mater 82:1–8 (December 2017)
14. Geary C, Birkinshaw C, Jones E (2008) Characterisation of Bionate polycarbonate polyurethanes for orthopaedic applications. J Mater Sci: Mater Med 19(11), 3355–3363. https://doi.org/10.1007/s10856-008-3472-8

A Research on the Cold Closed-Die Forging of Bevel Gears

Pham Quang Trung, Nguyen Hoang Dung, and Nguyen Nhat Minh

1 Introduction

Precision forging is defined as a flashless net-shape or near-net-shape forging operation using closed dies which generates high-quality parts concerning surface quality and dimensional accuracy [1, 2]. The products of precision forgings are used directly without any additional machining operations [2]. A number of researchers have reported on the precision forging of bevel gears [3–7]. These researches showed that the advantages of the precision forging technology are the high-dimensional precision of gears, less time to make gears, the increase in the mechanical properties of gears, simple operation, and simple equipment [3–7]. On the other hand, the research indicated the disadvantages are the complexity of the die structure, the difficulty to operate in hot forging, the necessity of having the equipment to make large force, and the suitability to produce small gears [3–7]. In 2013, Kang et al. [3] studied the cold closed-die forging of planetary bevel gears. In addition, the FE simulation researches on the cold forging of gears were also reported by Zhuang et al. [4]. Although the precision cold forging process has been used for many years, however, data on bevel gears forging process for lead materials under different conditions for simulation and experimental research are not readily available.

In this study, the ability of the fulfill materials in the closed-die process in precision forging bevel gears under different conditions has been systematically investigated in simulation and experimental works. Firstly, the simulation model is developed to simulate the forging process. Then, the mold is designed, manufactured, and assembled according to the simulation model. The experimental forging

P. Q. Trung (✉) · N. H. Dung · N. N. Minh
Ho Chi Minh City University of Technology, Ho Chi Minh City, Vietnam
e-mail: quangtrung@hcmut.edu.vn

P. Q. Trung · N. H. Dung · N. N. Minh
Vietnam National University, Ho Chi Minh City, Vietnam

R. K. Agarwal (ed.), *Recent Advances in Manufacturing Engineering and Processes*,
Lecture Notes in Mechanical Engineering,
https://doi.org/10.1007/978-981-16-3934-0_4

process is conducted, and the results are used to verify the numerical study. Finally, the simulation model after verified by the experiment is used to apply to another forging application.

2 Simulation Forming Process in Cold Forging

Within the scope of the study, pure lead is selected as a material for the simulation. Lead at room temperature is often similar to steel at the temperature of 1100 °C. Moreover, due to the good plastic deformation of lead, it is easy to conduct experiments at room temperature without having to heat the workpiece and dies like forging steel. The basic parameters of the chosen bevel gear are as a module $m = 2$, number of teeth $Z = 18$, reference cone angle $\varphi = 45°$, coefficient of correction $\xi = 0$, precision class IT = 8.

Deform 3D software is employed to simulate the forging process of bevel gear. This simulation determines the deformed state of the workpiece in the die and the force needed to operate. The geometric model for bevel gear in this study consists of punch, counterface, and upper and lower dies (Fig. 1a). The components are set according to the geometry of the bevel gear with the original workpiece volume approximating the volume of the component with the volume error rate within ±0.5 to ±1%.

The original workpiece originally meshes with a node number of 100,000 nodes. The boundary conditions of the simulation model are set as follows: the original workpiece's temperature for cold deformation is 25 °C. The temperature of the die is set to the same temperature as 25 °C. The punch and the dies are considered to be absolutely rigid, and therefore, they are not deformed during the forging process. The stroke of the punch is limited to a maximum of 23 mm in order to avoid pressing the counterface and the die directly to each other, and the movement of the punch is operated at 10 mm/min.

Fig. 1 Forging process: **a** FEM model. **b** Experimental study

(a) FEM model (b) Experimental

3 Experimental Study

After operating simulation and making die detail drawings, die sets are manufactured according to design technical requirements (Fig. 1b). American Instron machine (5980 Series Universal Testing System) with a forging force of 200 tons is employed to experiment. Table 1 shows the parameters of the experimental study.

4 Results and Discussion

The results of the process of forming gears when materials are forged into the die are presented in Table 2. The results show that when the stroke of the forging process is 10 mm, the workpiece begins to deform on both faces. When the stroke is 12 mm, the workpiece begins to form teeth. When the forging stroke is 15, 16, and 17 mm, the workpiece has a more clearly shaped gear and tooth profile. When the forging stroke is 18 mm, the material has filled the mold in key positions but has still not completely reached the tooth crowns. When the pressing stroke is 20.2 mm, the mold is fully filled and the tooth profile is shaped completely. Comparing this result with the research of Jin et al. [5], Song et al. [6], and Zhuang et al. [7] shows a good agreement of the photographs of cold-forged shapes of the bevel gear according to variation of punch strokes.

However, when the stroke exceeds 20.2 mm, a flash appears in the root. At this time, the material is filled into the die because the punch is still moving down, so the material is forced to flow out to the outside of the die groove (where the die parts are assembled together to make flash). At this time, the forging force on the machine increases significantly because of the forced metal flowing into these very small die grooves. Thus, in this experiment, the maximum stroke of the forged punch is determined to be 20.2 mm. It is concluded that the flash on the gear was created by the stretch of bolts according to a slightly larger punch stroke than the requirement [6].

Experimental results also show that when the punch speed increases from 10 to 15 mm/min, a better filling profile is received, but there is still flash in the root.

Table 1 Selected experimental parameters of forging bevel gear process

Sample	Stroke (mm)	Speed (mm/min)	Sample	Stroke (mm)	Speed (mm/min)
No. 1	0	0	No. 8	19	10
No. 2	10	10	No. 9	20	10
No. 3	12	10	No. 10	20.2	10
No. 4	15	10	No. 11	20.5	10
No. 5	16	10	No. 12	20.2	15
No. 6	17	10	No. 13	20.2	20
No. 7	18	10	No. 14	20.2	25

Table 2 Comparison of simulation and experimental results

Simulation	Exp.	Simulation	Exp.	Simulation	Exp.

No. 1	No. 1	No. 2	No. 2	No. 3	No. 3

No. 4	No. 4	No. 5	No. 5	No. 6	No. 6

No. 7	No. 7	No. 8	No. 8	No. 9	No. 9

No. 10	No. 10	No. 11	No. 11	No. 12	No. 12

No. 13	No. 13	No.14	No.14		

When the punch speed increases to 25 mm/min, the flash in the root of the tooth appears more. Consequently, at a speed of 10 mm/min, a 20.2 mm forging stroke is optimal. The process of metal filling in closed forging dies studied on simulations and on the experiment is showed in Table 2.

Comparing the simulation results and the experimental results, both studies show similar results in the shape of teeth when filling in die. Specifically, when the forging stroke is 10 mm, it is only possible to shape two technology holes when the stroke is 15 mm, the tooth profile begins to form and when the stroke is 20.2 mm, the tooth profile is shaped completely. The results provide more evidence on the accuracy and feasibility of gear fabrication by using the precision closed-die forging [5–7].

Figure 2a, b presents the loading effect according to the forging stroke in experimental and simulation works. Figure 2 shows what happens when the stroke increases gradually from 0 to 20.2 mm. This graph shows that when the forging stroke is from 0 to 10 mm, the force increases gradually but is small because this has only formed two small holes at the ends of the gear. Next, a significant increase in force is seen in the graph when the forging progressively increases in accordance

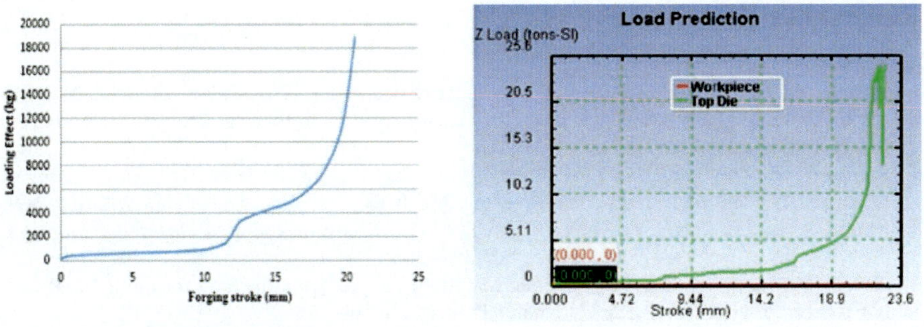

(a) Experiment study

(b) Simulation study

Fig. 2 Diagram of loading effect according to the forging stroke

with the process of forging metal into the die, the force is increasing because the metal begins to squeeze into difficult positions in the closed die. When the stroke is at the 20.2 positions, the metal has filled the die completely. If the forging continues, the forging force increases significantly because the metal escapes at the gap positions of the die parts. A comparison of the results in both figures shows that experimental and simulation works have similar results in the increment of loading when the stroke increases. A similar conclusion also was found in the report of Song et al. [6], in which the forming load simulated from the simulation matched well with that of experiments in the research of low carbon steel AISI1020.

5 Conclusions

The simulation and experiment study on the fulfill materials in the closed-die process in precision forging bevel gear under different conditions has been investigated. The following conclusions can be drawn from the obtained results:

- The comparison between the simulation and experimental works presents similar results in the shape of teeth when filling in the mold.
- The increase in the forging stroke results in the increment of the forging force, which reaches a value of 19 and 20 tons when forging stroke is over 20.2 mm in the simulation and experiment works, respectively. It is seen that the force between simulation and experiment is not significantly different, and the ability to fill the cavity is good.
- Finally, the speed of stroke in the range of 10–25 mm/min does not affect the forming process in both studies.

Acknowledgements This research is funded by Ho Chi Minh City University of Technology—VNU-HCM, under grant number T-CK-2019-04.

References

1. Fuentes A, Iserte JL, Gonzalez-Perez I, Sanchez-Marin FT (2011) Computerized design of advanced straight and skew bevel gears produced by precision forging. Comput Methods Appl Mech Eng 200:2363–2377
2. Marini D, Cunningham D, Corney JR (2017) Near net shape manufacturing of metal: a review of approaches and their evolutions. Proc Inst Mech Eng Part B: J Eng Manuf 232:650–669
3. Kang F, Yang E, Wang Y, Chen Q, Shu D (2013) Study on the cold closed-die forging of planetary bevel gears. Adv Mater Res 803:321–325
4. Zhuang W, Han X, Hua L, Xu M, Chen M (2019) FE prediction method for tooth variation in hot forging of spur bevel gears. J Manuf Process 38:244–255
5. Jin J-S, Xia J-C, Wang X-Y, Hu G-A, Liu H (2009) Die design for cold precision forging of bevel gear based on finite element method. J Cent South Univ Technol 16:546–551
6. Song JH, Im YT (2007) Process design for closed-die forging of bevel gear by finite element analyses. J Mater Process Technol 192–193:1–7
7. Zhuang W, Hua L, Han X, Zheng F (2017) Design and hot forging manufacturing of non-circular spur bevel gear. Int J Mech Sci 133:129–146

Bead Width Prediction in Laser Wire Additive Manufacturing Process

Natago Guilé Mbodj and Peter Plapper

1 Introduction

In LWAM, a laser beam melts a metal wire to form a weld seam [1]. The final part's accuracy is influenced by several factors such as the thermal history, bead geometry (height and width), travel speed, power [2–6]. Several studies show that the most important parameter affecting deposition is the power [7, 8]. In [2], Yakout et al. showed that increasing the power augments the focus area so increases the bead width. In [9], Heralić et al. varied the laser power and notice a melt pool width change, thus a fluctuation of bead width geometry.

The advantages of using a laser beam are the following [10]:

With a laser beam, the power generated can be adjusted to a small area or larger area if needed.
The laser beam is easier to control.

Another advantage is the adjustability of the stand-off distance preventing a collision between the workpiece and the deposition head.

However, laser beam presents disadvantages such as difficulty to work with reflective materials (copper, gold) and the expensiveness compared to an electric arc or electron beams.

In the literature, we found some works where the other process parameters also influence the bead geometry [9, 11–13].

In [11], Plangger et al. theoretical study's showed the influence of the wire speed velocity and the travel speed on the bead geometry. Further, Plangger et al. encouraged to estimate bead geometry before choosing the wire feed speed and travel speed for a deposition process. Besides the main process parameters, temperature, as an example, influences the bead geometry formation. Bi et al. [14]

N. G. Mbodj (✉) · P. Plapper
University of Luxembourg, 6, rue Coudenhove-Kalergi, 1359 Esch-sur-Alzette, Luxembourg
e-mail: natago.mbodj@uni.lu

R. K. Agarwal (ed.), *Recent Advances in Manufacturing Engineering and Processes*,
Lecture Notes in Mechanical Engineering,
https://doi.org/10.1007/978-981-16-3934-0_5

study's confirms this influence and raises an inter-dependency between the temperature and the main process parameters.

The wrong combination of parameters generally leads to porosity in the final part, which is the main problem in metal additive manufacturing [12]. For this reason, some of the research is oriented along the vein of the bead geometry prediction process. In [15], Xiong et al. applied a second-order regression and neural network to predict the weld bead geometry. The results show that the neural network can predict bead geometry more accurately than the second-order regression model. However, this prediction is applied for a single bead, so it is difficult to judge the model's robustness in the case of multi-bead deposition. In [16], Dhas and Kumanan proposed a technique called adaptive neuro-fuzzy inference system (ANFIS) to predict the bead width for a set of process parameters in a submerged arc welding (SAW) process. A multiple regression model is used to train the data. The results show less model error compared to artificial neural network (ANN) model. However, both ANN and Dhas et al.'s methods are not proven yet in the AM process because of many other physical parameters interfering in the deposition process, making the process nonlinear and unpredictable.

This paper proposes a bead width prediction using a regression algorithm. In the experiments, the laser power, wire feed rate, and advanced speed of the robot are kept constant, as given in Table 1. The regression algorithms are coded in Python. The results show an *n*th root evolution of the bead width across layers. Our study reveals the possibility to perform a reliable and stable process with fewer defects and high part accuracy using the prediction trend to find the right combination of process parameters. In a separate paper, height prediction of layer thickness is performed. The remainder of this paper is structured as follows: Sect. 2 describes the experimental method. Section 3 explains the bead width prediction model. Section 4 contains the results, and Sect. 5 summarizes the paper with a conclusion.

Nomenclature
$F(x)$ bead width prediction function
i index for the training set $(0, 1, 2, 3, \ldots, n)$
j index of increment $(0, 1, 2, 3, \ldots, n)$
$h(x)$ predictor or hypothesis
$x^{(i)}$ number of layers (input variable)
$y^{(i)}$ bead width (output variable)
$(x^{(i)}, y^{(i)})$ training example
X input space
Y output space
θ_j parameter of the regression function
m number of training examples (layers)
$J(\theta)$ cost function
α learning rate

Table 1 Process parameters selected for the regression experiments

Selected trials (samples)	Laser power (kW)	Travel speed (m/min)	Wire feed rate (m/min)
1	1.6	0.6	2
2	1.6	0.3	2.2
3	2	0.9	2.2

2 Experimental Method

Experiments are performed to see the influence of parameters on the bead geometry. The main process parameters varied in the experiments are the laser power, the advanced speed, and the wire feeder rate. In preparatory experiments, the best process parameters were determined, as given in Table 1. One typical cross-section as shown in Fig. 1 is generated during the preparatory experiments where the requirement is to obtain a smooth deposition with less visible stacking layers. The wire is orthogonal to the deposition plan and not preheated.

The method is to deposit ten layers of beads with varying process parameters values. Several trials were performed; then, trials with good results (straight beads), as shown in Fig. 1, are principal candidates for the regression algorithm.

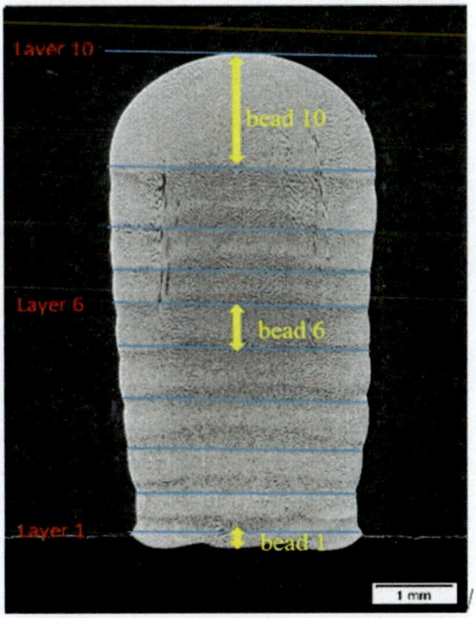

Fig. 1 Ten deposited beads from a right combination of process parameters

3 Bead Width Prediction Model

Learning from data is common now in engineering. Learning algorithms' goal is to extract useful information from collected data [17]. In the paper, a regression algorithm is employed like the following: Given a set of deposited beads, bead widths across layers are predicted. The prediction model takes as input the training examples $(x^{(i)}, y^{(i)})$; $i = 1$ to m, $m = \mathit{final\ layer}$ extracted from Fig. 1. A predictive function $F(x)$ is received as output. The function relates the evolution of the bead width over layers. Pre-experiments proved that with multiple layer deposition of beads, the general trend shows that each superposed bead is larger than the previous. The prediction algorithm is implemented in Python language and has the following parts: First, a mapping function is defined as $b \times X^a$. $a = \frac{1}{c}; c = 2, 3, \ldots, \infty$, defines the vertical stretch or shrink of the curve and b the starting point of the curve. Then, the training set is plotted and data linearized with a natural logarithm. After, a linear regression algorithm is applied to find gradient descent values. Finally, an inverse transformation gives the parameters used to compose the bead width prediction function $F(x)$.

The process of the bead width prediction is shown in Fig. 2. The goal is to create a "good" predictor function $h : X \rightarrow Y$ so that $h(x)$ predicts the bead width growth and where $X = Y = R$. The prediction model is composed of the following parts [18]:

A database stores the training examples. $x^{(i)}$ is the "input" variables (number of layers) and, $y^{(i)}$ is the target variable (predicted bead width). $(x^{(i)}, y^{(i)})$ is a training example.

A common choice is to approximate y as a linear function of s after the linearization is shown in Eq. (1).

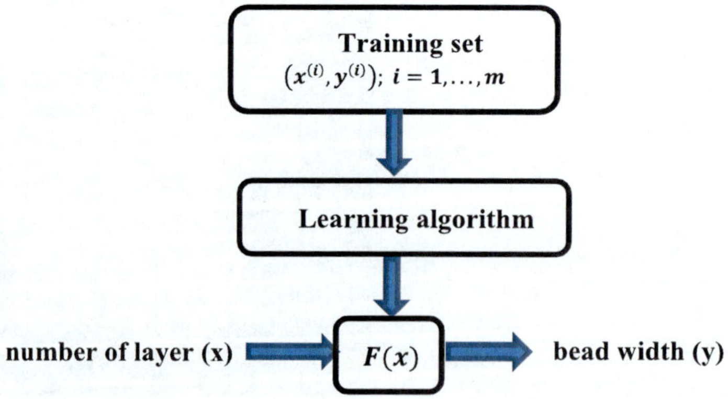

Fig. 2 Bead width prediction process

$$h(x) = \theta_0 + \theta_1 x \quad (1)$$

The linear function $h(x)$ fits the parameters $\theta_i s$ for a list of training examples $\{(x^{(i)}, y^{(i)}); i = 1 \text{ to } m\}$. Then, the cost function in (2) measures how clos $h(x^{(i)})$'s are to the corresponding $y^{(i)}$.

$$J(\theta) = \frac{1}{2}\sum_{i=1}^{m}\left(h_\theta\left(x^{(i)}\right) - y^{(i)}\right)^2 \quad (2)$$

After computing the cost function $J(\theta)$, a search algorithm called gradient descent in (3) is used to minimize $J(\theta)$.

$$\theta_j := \theta_j - \alpha\frac{\partial}{\partial\theta_j}J(\theta) <== > \theta_j := \theta_j + \alpha\sum_{i=1}^{m}(y^{(i)} - h_\theta(x^{(i)}))x_j^{(i)} (\text{for every } j) \quad (3)$$

The number of iteration and learning rate are selected in a manner for the algorithm to converge to the best values of gradient descents. Tuning experiments helped to choose the values. In this paper, a number of iteration of 2000 and a learning rate if 0.03 were used.

After the gradient descent values θ_0 and θ_1 that minimize the cost function are obtained, θ_s are run to obtain the bead width prediction function. The function $F(x)$ predicts the bead width for a given number of layers and shows similar trends for all performed experiments.

The prediction process is summarized in the flowchart shown in Fig. 3. If the fitting curve satisfies the inputs/outputs function mapping, then the prediction function $F(x)$ showing the bead width evolution across layers can be collected. If not, the parameters a and b of the mapping function $b \times X^a$ and the numerical hyperparameters of the regression algorithm (learning rate and number of iterations) must be tuned.

4 Results and Discussion

The result of the prediction task is shown in Fig. 4. The bead width increases in a nth root function trend for the given range of layers (1 to 10) as shown in Fig. 4a. The results show that the prediction curve can be designed and adapted to a given training set to relate the bead width evolution across layers. In this work, few trials allowed to define the mapping function parameters and the regression algorithm selection to fit any given training set where the bead width evolution trend is respected. As a validation, the prediction curve is extended to five layers as shown in Fig. 4b. The results show that from layer 11 to layer 15, the bead width change across layers (green rectangle) is very close to the prediction curve. To summarize,

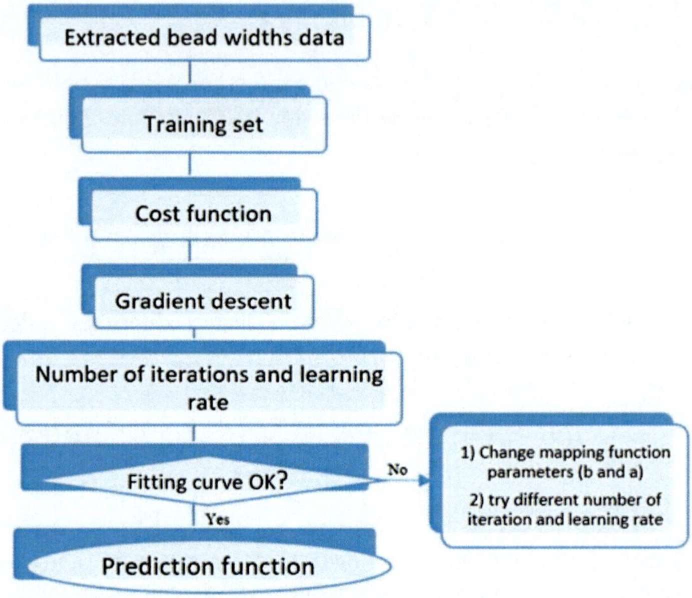

Fig. 3 Flowchart of the bead width prediction model

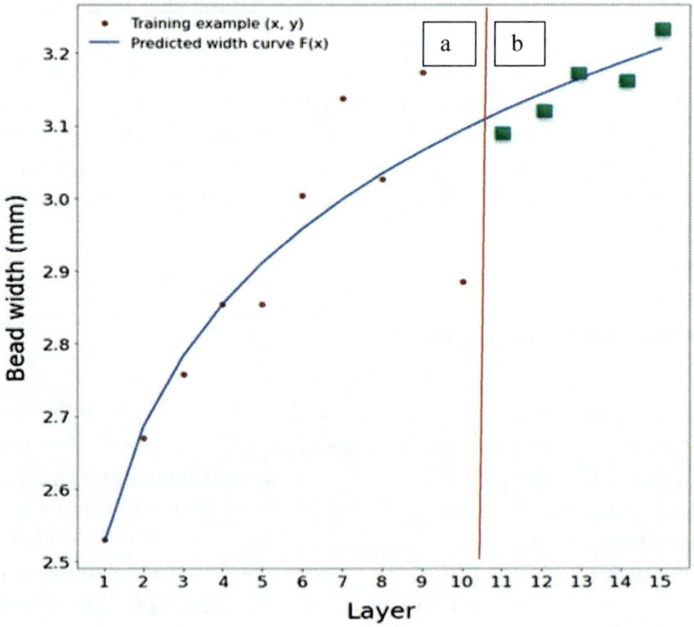

Fig. 4 Experiment trial with set of process parameters selected: **a** result, **b** validation

the proposed bead width prediction model offers design flexibility and is adaptable compared to ANN models, where the model performance depends on many trials and hidden layers. Also, the prediction curves do not perfectly fit the training set because the goal is not to overfit the training data but to approximate the training set to obtain a function correlating the bead width evolution across the layer. Finally, as a prerequisite, the bead width prediction technique needs to have some good deposition trials from which it applies the prediction model. In the future, other process parameters intervening in 3D printing process need to be included for a more robust model.

5 Conclusions

A bead width prediction model is developed using regression algorithms. The algorithm developed in Python receives the bead width data, predicts the bead width evolution, and confirms a profile trend for stable deposition. The model is flexible and can be adjusted to follow the trend of various nonlinear curves if needed.

The following conclusions are drawn:

1. Regression algorithms can be used for prediction tasks related to the additive manufacturing process but with supervision.
2. The existence of a curve trend correlates the process parameters to the bead width growth across layers.
3. The approach can allow a faster parameter selection in the future and could be a time gain for deposition with other materials and even different shapes.
4. A relation exists between the process parameters, the bead width, and the layer height thickness which is investigated in another paper.

Acknowledgements This work was supported by the Interreg V-A Grande Région "Fabrication Additive par Dépôt de Fil" (Fafil) project. The authors would like to thank the Institut De Soudure Industrie—Yutz for providing the data.

References

1. Li Y, Huang X, Liu Y, Peng H, Azer M (2005) Laser net shape manufacturing of metallic materials with CO_2 and fiber laser. In: ICALEO congress proceedings, Miami, FL, USA, pp 320–325
2. Yakout M, Cadamuro A, Elbestawi MA, Veldhuis SC (2017) The selection of process parameters in additive manufacturing for aerospace alloys. Int J Adv Manuf Technol 92:2081–2098
3. Apps R, Gourd L, Nelson K (1963) Effect of welding variables upon bead shape and size in submerged-arc welding. Weld Metal Fab 453–445

4. Wang C, Bai H, Ren C, Fang X, Lu B (2020)A comprehensive prediction model of bead geometry in wire and arc additive manufacturing. Journal of physics: conference series, volume 1624, computer modeling and simulation technology
5. Dinovitzer M, Chen X, Laliberte J, Huang X, Frei H (2019) Effect of wire and arc additive manufacturing (WAAM) process parameters on bead geometry and microstructure. Addit Manuf 26:138–146
6. Mok SH, Bi G, Folkes J, Pashby I (2008) Deposition of Ti–6Al–4V using a high power diode laser and wire, part i: investigation on the process characteristics. Surf Coat Technol 202(16): 3933–3939
7. Heralić A, Christiansson AK, Ottosson M, Lennartson B (2010) Increased stability in laser metal wire deposition through feedback from optical measurements. Opt Lasers Eng 48(4): 478–485
8. Grünenwald S, Unt A, Salminen A (2018) Investigation of the influence of welding parameters on the weld geometry when welding structural steel with oscillated high-power laser beam. Procedia CIRP 74:461–465
9. Heralić A, Christiansson AK, Hurtig K, Ottosson M, Lennartson B (2008) Control design for automation of robotized laser metal-wire deposition. IFAC Proc Vol 41(2):14785–14791
10. J. Blackburn (2012) Laser welding of metals for aerospace and other applications. Woodhead publishing series in welding and other joining technologies, pp 75–108
11. Plangger J, Schabhüttl P, Vuherer T, Enzinger N (2019) CMT additive manufacturing of a high strength steel alloy for application in crane construction. Metals Open Access Metall J 9(6):650
12. Song L, Singh VB, Dutta B, Mazumder J (2012) Control of melt pool temperature and deposition height during direct metal deposition process. Int J Adv Manuf Technol 58:247–256
13. Pathak D, Singh RP, Gaur S, Balu V (2020) To study the influence of process parameters on weld bead geometry in shielded metal arc welding. Mater Today Proc
14. Bi G, Gasser A, Wissenbach K, Drenker A, Poprawe R (2006) Identification and qualification of temperature signal for monitoring and control in laser cladding. Opt Lasers Eng 44(12): 1348–1359
15. Xiong J, Zhang G, Hu J, Wu L (2014) Bead geometry prediction for robotic GMAW-based rapid manufacturing through a neural network and a second-order regression analysis. J Intell Manuf 25:157–163
16. Dhas JER, Kumanan S (2007) ANFIS for prediction of weld bead width in a submerged arc welding process. Eng J Sci Ind Res
17. Doan T, Kalita J (2015) Selecting machine learning algorithms using regression models. In: ICDMW'15: proceedings of the 2015 IEEE (ICDMW), pp 1498–1505
18. Andrew N CS229: machine learning, cs229.stanford.edu [Online accessed: 11-06-20]

Investigation the Ultrasonic Injection Molding of Polyamide 6

Thanh-Hai Nguyen, Tran Long Vu, Vo Van Duy Phong, Tra Ngoc Tien Dat, Anh Van Nguyen, Pham Son Minh, and Nguyen Huu Loc

1 Introduction

With the development of science and technology in recent years, the trend toward miniature/micro thermoplastic parts has increased rapidly. The most common and widely used technologies to produce micro plastic parts are injection molding and hot embossing. However, these technologies still face many challenges. In small and single production, the injection molding method has a noticeable disadvantage. Because the plastic parts are of a micro size, the amount of material wasted after the process can be huge compared to the actual products. This technology also requires very high pressure and external heat to melt and fill plastic into micro molds, resulting in large and expensive equipment. In hot embossing, the mold needs to be heated, the molding period is long and the products are hard to demold. Therefore, researchers are constantly working to find a method for miniature/micro molding plastic parts that can solve the problems listed above.

Ultrasonic technology has been widely used in many different fields in plastic processing, especially in thermoplastic welding [1] and nonwoven fabrics [2], because of its good characteristics; it is clean, inexpensive, energy efficient, pro-

T.-H. Nguyen (✉) · T. L. Vu · V. Van Duy Phong · T. N. T. Dat · N. H. Loc
Faculty of Mechanical Engineering, Ho Chi Minh City University of Technology (HCMUT), 268 Ly Thuong Kiet Street, District 10, Ho Chi Minh City 700000, Vietnam
e-mail: haint@hcmut.edu.vn

T.-H. Nguyen · N. H. Loc
Viet Nam National University Ho Chi Minh City, Linh Trung Ward, Thu Duc District, Ho Chi Minh City 700000, Vietnam

A. Van Nguyen
Research and Development Department, Murata Welding Laboratory, Osaka 532-0012, Japan

P. S. Minh
Ho Chi Minh City University of Technology and Education, Ho Chi Minh City, Vietnam

R. K. Agarwal (ed.), *Recent Advances in Manufacturing Engineering and Processes*, Lecture Notes in Mechanical Engineering,
https://doi.org/10.1007/978-981-16-3934-0_6

duces little waste and has a fast process. While injection molding requires external heat to melt plastic before the molding process and the mold needs to be heated in hot embossing, ultrasonic molding uses heat derived from the interfacial friction effect between the plastic powder and the viscoelastic dissipation of the polymer to melt the thermoplastic within the molding process [3, 4]. As a result, an innovative molding method called ultrasonic micro-injection molding (UMIM) has been researched and developed in recent years. This technology uses heat derived from the interfacial friction effect between plastic pellets and the viscoelastic dissipation of the polymer to melt the thermoplastic within the molding process [3]. The method benefits from high economic efficiency for small and medium production [4]. Ultrasound increases the temperature of plastic particles very quickly, making the product cycle short, and so, the plastic is only kept at high temperatures for a short time to avoid damaging the products. Micheali et al. were the first to apply ultrasound in the micro-injection molding process in 2002 [5]. Their experiments showed that ultrasonic energy can be used to melt a sufficiently small amount of plastic for micro-injection molding and also reduces the cycle time of the process [5, 6]. Chena et al. proved that by increasing the ultrasonic energy, the molecular weight of plastic is reduced and the molecular weight distribution narrowed down, reducing the viscosity of a polymer and thereby improving the molding properties of some materials [7]. Liang et al. employed ultra-high molecular weight polyethylene (UHMWPE) powders to fabricate micro gears; their research indicated that, compared to micro-injection molding, the micro ultrasonic polymer molding process directly plasticizes the polymer in the micro cavity, greatly reducing the cycle time, and prevents the formation of weld lines and other defects in micro plastic parts [8]. Jiang et al. analyzed the relationship between the ultrasonic and polymer plasticization speeds as well as the internal mechanisms and found that the ultrasonic plasticization speed of the polymer increases with the ultrasonic supply voltage and plasticization pressure [9]. When increasing the ultrasonic time and pressure, the flashes gradually thin out and automatically separate from the micro parts after exceeding the critical value [10]. Polyamide 6 (PA6) is a semi crystalline polymer which is widely used in injection molding, and it is a good candidate for micro-injection molding because of its high mechanical performance, good processability and low melt viscosity [11]. They can also be molded at about 80 °F (27 °C) lower temperatures with less mold shrinkage because they are slightly less crystalline. Depending on the crystallization conditions, PA6 may develop two different crystalline phases—the α-phase and γ-phase—which convey different properties to the solid polymer [12, 13]. Recently, researchers have reported that PA6 can be utilized to produce gears, composites or micro parts by injection molding [14–16]. However, the tensile strength and the elongation were low in comparison with the raw material [17–19]. Because of this, a novel approach for molding PA6 is needed to improve the efficiency and the quality of molding products using the PA6 material. In this study, two ultrasonic horns are designed by finite element method, fabricated using aluminum 7075 alloy, and applied for producing the tensile samples with PA 6 materials. Some major processing parameters and its interaction such as pressure, ultrasonic applying time, holding time and molding temperature were also investigated. The results show that tensile samples are fully generated by ultrasonic heat from the raw PA 6 pellets. In addition, the pressure and mold temperature parameters affected significantly to the ultrasonic miniature injection molding for the tensile samples.

2 Experiment

The process of ultrasonic injection molding is expressed in Fig. 1, carrying out in some main stages. Firstly, PA 6 pellets are inserted into the plastic chamber, Fig. 1a. In the second stage, the ultrasonic horn moves down and activates the ultrasonic wave just right before it comes in contact with the raw pellets. Those are then melted and blended with each other, Fig. 1b. In the third stage, all pellets are completely melted and began to fill the mold cavity, Fig. 1c. In the final stage, the material completely fills the mold cavity and starts to cool down, Fig. 1d. During the cooling period, the position of the ultrasonic horn still remains constantly, always at the lowest position and maintaining pressure on the melted plastics. Finally, the ultrasonic horn returns to the original position and the mold is separated to extract the sample.

In this research, ultrasonic horns of a half wavelength and full wavelength are designed and manufactured, Fig. 2. Horn is one of the important parts that significantly affecting to the injection mold. The ultrasonic horn is the vibrating transmitter from the ultrasonic generator directly to plastic granules and needs to be designed in accordance with the specific frequency (20 kHz) and working diameter (10 mm). Therefore, it is designed, simulated and fabricated based on finite element method. At first experiment, a half wavelength was designed by Abaqus software and fabricated using aluminum 7075 alloy [1, 2]. In this case, the half wavelength horn gives a more compact shape compared to the full wavelength horn. It also generates more concentrated stresses at the zero-amplitude position, as seen in the red area in upper part of Fig. 2a. When mounting the fabricated horn with a

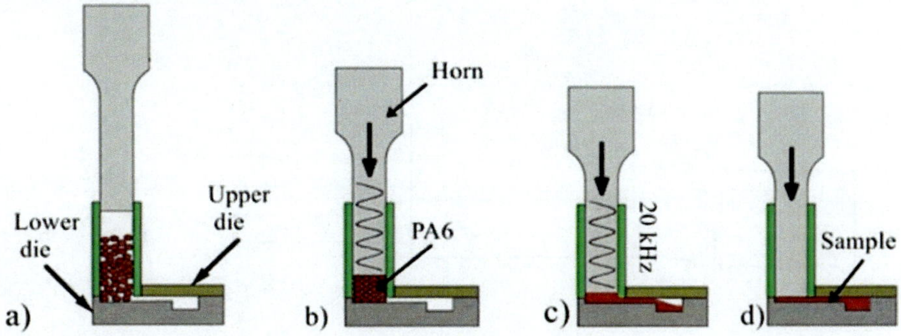

Fig. 1 Principle diagram of ultrasonic miniature injection molding

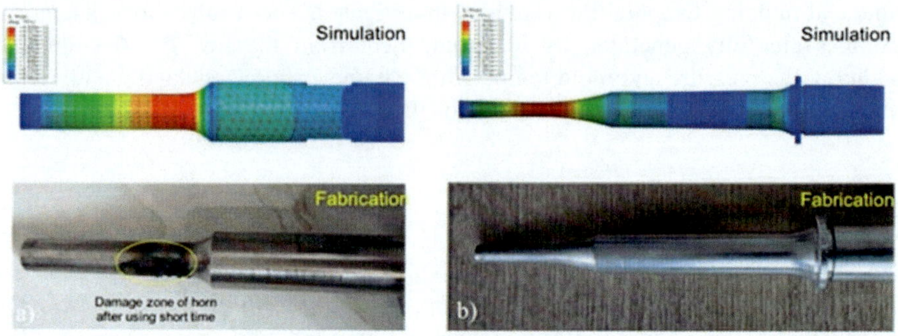

Fig. 2 Simulation and fabrication of: **a** the half wave length and **b** full wavelength horns

transducer at high ultrasonic intensity, the horn is overheating at the concentrated stresses and destroys the horn, Fig. 2a. In order to overcome this issue, a re-design of ultrasonic horn must be considered. Therefore, a horn with a full wavelength was employed, Fig. 3. The horn length is longer but the concentrated stresses are reduced rapidly, seen in upper part of Fig. 2b. The horn was successfully designed and fabricated using aluminum 7075 alloy, as exhibited in lower part of Fig. 2b. Therefore, this horn is ready for applying in all experiments.

The simple tensile sample is utilized as a miniature part, Fig. 4. The injection mold consists of two main parts: the upper die and lower die. Both dies are aligned and connected precisely thanks to two locating pins. The plastic chamber is the place where the plastic pellets are occupied, melted and filled to the cavity. The gap between the plastic chamber and the ultrasonic horn is small enough that the melted material will not leak as press. The mold material was aluminum alloy, grade A6061. The mold was designed by AutoCAD 3D software and fabricated with a basic size of 200 mm in length, 90 mm in width and 106 mm in height, Fig. 5.

The material used in this project is Polyamide 6 (PA6) of non-reinforced, injection molding grade produced by Lanxess Energizing Chemistry company. The ultrasonic molding equipment utilizes in this work includes an ultrasonic generator,

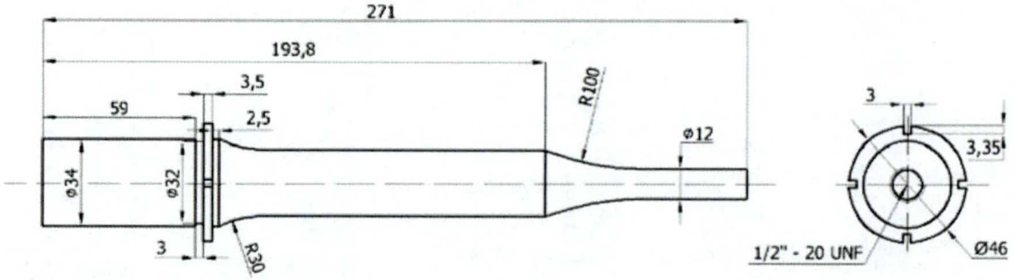

Fig. 3 Drawing of full wave length of ultrasonic horn

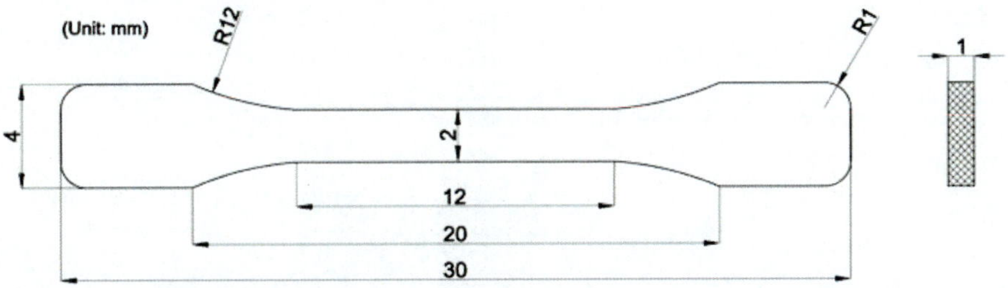

Fig. 4 Design of ultrasonic-assisted tensile strength sample

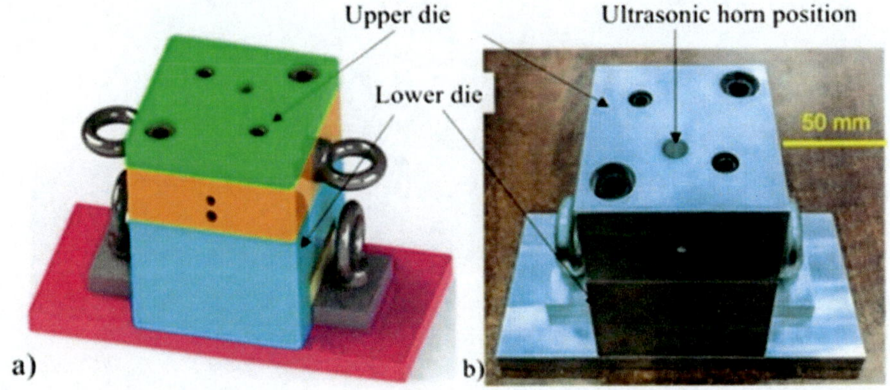

Fig. 5 Ultrasonic-assisted injection mold: **a** CAD model and **b** fabrication

heating resistor and sensor, ultrasonic lateral air cylinder, electrical control box, ultrasonic horn and ultrasonic micro-injection mold, a product of Viet Nam Ultrasonic Equipment Company, Vietnam, Fig. 6.

3 Results and Discussion

In this experiment, the main parameters of pressure, ultrasonic applying time, holding time and mold temperature were set at 0.2 MPa, 2 s, 5 s, 60 °C, respectively. A typical sample of ultrasonic-assisted injection molding of PA 6 was obtained, Fig. 7. The PA 6 pellets melted and filled fully to the cavity under ultrasonic heat, Fig. 7a. In other cases, many voids occur at the end of the tensile samples, Fig. 7b, c. There are some causes for these defects. Ultrasonic heat generates insufficient for melting all the pellets. The air pressure reduces gradually during the ultrasonic excitation or ultrasonic applying time is smaller than expected.

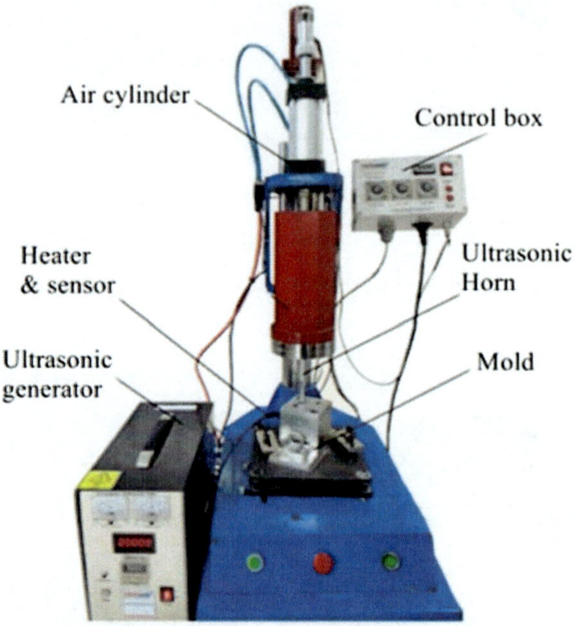

Fig. 6 Experiment setup for ultrasonic injection molding

Fig. 7 Typical sample of ultrasonic-assisted injection molding

In order to analyze the influence of individual parameters and the interactive parameters, a series of experiments were conducted and analyzed on the tensile strength of plastic parts. The pressure, ultrasonic applying time, holding time and mold temperature were set at two levels: 0.2–0.3 MPa, 2–3 s, 5–10 s, 60–80 °C, respectively. Totally, 32 samples are produced base on these setting parameters. A Pareto chart was constructed by Minitab software using an analysis of variance (ANOVA), Fig. 8. The red dashed line in the graph, setting at 2.12, is the reference line. The Pareto plot shows that the mold temperature (D) and pressure (A) are the two factors significantly affected at the 95% confidence level. However, the mold temperature has slightly more influence than pressure. Conversely, the ultrasonic applying time (B), interaction between pressure and mold temperature (AD),

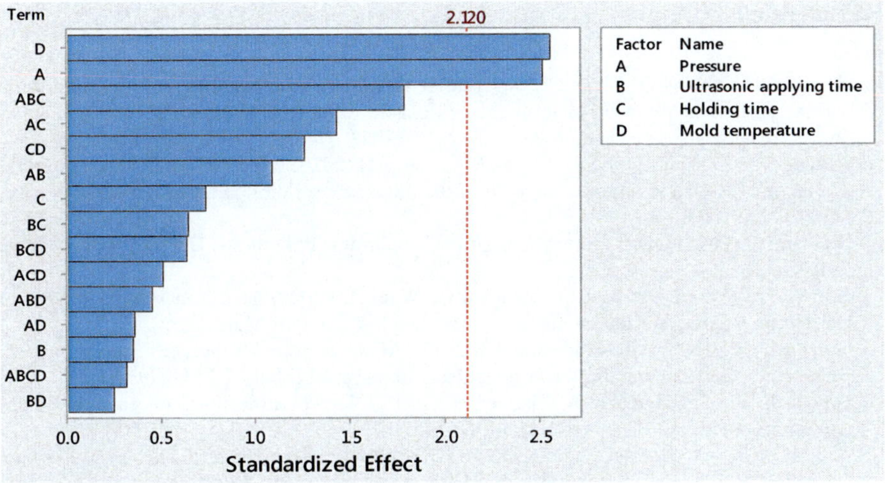

Fig. 8 Pareto chart of the standardized effect on tensile strength

interaction between ultrasonic time and mold temperature (BD) have insignificant effects on the response. Combining three parameters (ABC), pressure–ultrasonic applying time–holding time had the most influence. However, this interaction is still located on the left side of the red dashed line.

4 Conclusions

In this work, ultrasonic injection mold using PA6 was investigated on the tensile sample. The ultrasonic horns were designed by finite element method and fabricated using aluminum 7075 alloy in cases of half wave length and full wave length. Half wave length horn was broken because of high concentrated stresses and full wave length horn was utilized for the experiment. Four major factors including pressure, ultrasonic applying time, holding time and mold temperature were investigated during the experiment. The typical tensile sample was obtained clearly. The Pareto plot shows that the pressure and mold temperature are the two factors that have the most influence on the tensile strength.

Acknowledgments This research is funded by Ho Chi Minh City University of Technology—VNU-HCM, under grant number T-CK-2020-02.

References

1. Nguyen TH, Quang QT, Tran CL, Nguyen HL (2017) Investigation the amplitude uniformity on the surface of the wide-blade ultrasonic plastic welding horn. In: 2017 IOP conference series: materials science and engineering 241:012023
2. Nguyen T-h, Thanh LQ, Loc NH, Huu MN, Van AN (2020) Effects of different roller profiles on the microstructure and peel strength of the ultrasonic welding joints of nonwoven fabrics. Appl Sci 10:4101
3. Farhan RS (2015) Ultrasonic welding of thermoplastics, PhD thesis, University of Sheffield, UK
4. Zhang ZB, Wang XD, Luo Y, Zhang ZQ, Wang L (2009) Study on heating process of ultrasonic welding for thermoplastics. J Thermoplast Compos Mater 22:6
5. Michaeli W, Kamps T, Hopmann C (2011) Manufacturing of polymer micro parts by ultrasonic plasticization and direct injection. Microsyst Technol 17:243–249
6. Michaeli W, Spennemann A, Gartner R (2002) New plastification concepts for micro-injection moulding. Microsyst Technol 8:55–57
7. Chena J, Chen Y, Huilin L, Lai SY, Jow J (2010) Physical and chemical effects of ultrasound vibration on polymer melt in extrusion. Ultrason Sonocemistry 17:66–71
8. Liang X, Wu X, Zheng K, Xu B, Wu S, Zhao H, Li B, Ruan S (2014) Micro ultrasonic powder molding for semi-crystalline polymers. J Micromech Microeng 24:045014
9. Jiang BJ, Hu JL, Li J, Liu XC (2012) Ultrasonic plastification speed of polymer and its influencing factors. J Central South Univ 19:380–383
10. Zeng K, Wu XY, Liang X, Xu B, Wang YT, Chen XQ, Cheng R, Luo F (2014) Process and properties of micro-ultrasonic powder molding with polypropylene. Int J Adv Manuf Technol 70:515–522
11. Zhang C (2018) Progress in semicrystalline heat-resistant polyamides. E-Polym 18:373–408
12. Wu Y, Huang A, Fan S, Liu Y, Liu X (2020) Crystal structure and mechanical properties of uniaxially stretched PA612/SiO2 films. Polymers 12:711
13. Liu Y, Cui L, Guan F, Gao Y, Hedin NE, Zhu L, Fong H (2007) Crystalline morphology and polymorphic phase transitions in electrospun nylon 6 nanofibers. Macromolecules 40
14. Zhao J, Mayes RH, Chen G, Xie H, Chan PS (2003) Effects of process parameters on the micro molding process. Polym Eng Sci 43:1542–1554
15. Tavcar J, Grkman G, Duhovnik J (2018) Accelerated lifetime testing of reinforced polymer gears. J Adv Mech Des Syst Manuf 12:1–12
16. Hakimian E, Sulong AB (2012) Analysis of warpage and shrinkage properties of injection-molded micro gears polymer composites using numerical simulations assisted by the Taguchi method. Mater Des 42:62–71
17. Chrzanowska E, Gierszewska M, Kujawa J, Raszkowska KA, Kujawski W (2018) Development and characterization of polyamide-supported Chitosan nanocomposite membranes for hydrophilic pervaporation. Polymers 10:868
18. Hooreweder BV, Moens D, Boomen R, Kruth JP, Sas P (2013) On the difference in material structure and fatigue properties of nylon specimens produced by injection molding and selective laser sintering. Polym Testing 32:972–981
19. Kunishima T, Kurokawa T, Arai H, Fridrice V, Kapa P (2020) Reactive extrusion mechanism, mechanical and tribological behavior of fiber reinforced polyamide 66 with added carbodiimide. Mater Des 188:108447

In-Situ Monitoring for Open Low-Cost 3d Printing

P. Minetola, M. S. Khandpur, L. Iuliano, F. Calignano, M. Galati, and L. Fontana

1 Introduction

Nowadays, additive manufacturing (AM) is complementary to traditional subtractive manufacturing in many industrial sectors. Increased design freedom is the driver for pursuing specific benefits and improvements in part performances for successful application of AM [1]. From an environmental perspective, the sustainability of AM and 3D printing has been investigated in the literature [2].

When compared to traditional manufacturing, AM offers the opportunity for in-situ monitoring the fabrication process to detect defects within the single layer. When an important defect is present, the user can decide to abort the part production, avoiding unnecessary material waste with related economic advantages in terms of time and cost. For industrial applications, in-situ monitoring plays a key role since it substitutes conventional non-destructive testing (NDT) methods that should be applied after part fabrication with an increase in cost and times. For in-situ layer-wise inspection, many techniques have been proposed in the literature [3–8].

Among AM technologies [9], the most widespread process is fused deposition modeling (FDM), which was initially patented by Stratasys company. FDM is based on the extrusion of a thermoplastic filament through a nozzle for the deposition of the single layer, after melting the material in a heated chamber. FDM is also more popularly renowned as 3D printing of fused filament fabrication (FFF), and the expiration of Stratasys patent has favored the diffusion of a great number of low-cost 3D printers.

Several possible defects can occur during 3D printing with detrimental effects on the structural or aesthetic quality of the product [10–12]. These defects can be

P. Minetola (✉) · M. S. Khandpur · L. Iuliano · F. Calignano · M. Galati · L. Fontana
Department of Management and Production Engineering (DIGEP), Politecnico di Torino, Corso Duca degli Abruzzi 24, 10129 Torino, Italy
e-mail: paolo.minetola@polito.it

R. K. Agarwal (ed.), *Recent Advances in Manufacturing Engineering and Processes*, Lecture Notes in Mechanical Engineering,
https://doi.org/10.1007/978-981-16-3934-0_7

monitored in each layer using computer image analysis for detecting (i) missing or (ii) unwanted filament deposition.

3D printers are sold in the market at a price starting from about 100€ in the case of kits that the users should assemble by themselves. In the design of the monitoring system, two aspects were considered. First, the additional sensors cannot cost more than the printer for the additional feature of in-situ monitoring not to be more expensive than the primary function of layer-wise fabrication. Secondly, the integration of automatic monitoring procedures requires the 3D printer code to be open for introducing an inspection phase at the end of the fabrication of every single layer.

In this paper, a low-cost in-situ monitoring system is implemented in an open 3D printer by the addition of a CCD camera for automatic detection of defects during part production using computer vision algorithms in Matlab. A brief description of the in-situ monitoring system and procedure is provided in the following section. The application of the monitoring process is presented in the third section. A summary of this work and its future implementation are presented in the conclusions.

2 Materials and Methods

The online monitoring system has been implemented with a USB camera model SV-USBFHD06H-SFV by Svpro. This camera was selected because of the best tradeoff between resolution and cost. The camera has a Sony IMX322 sensor and comes with a 5–50 mm varifocal lens of 1/2.9″ size. It has a resolution of 2 Megapixels (1920 $\times$ 1080 pixels) and costs about 85 euros. The focus can be manually adjusted depending on the distance between the installation position and the print bed. The sensor sensitivity allows the camera to correctly operate in scarce light since the minimum illumination should be 0.01 lx.

For the implementation of the in-situ monitoring system, an A4v3 3D printer by 3ntr was chosen. The printer can print parts using three extruders that are fed with 2.85 mm filaments. The machine comes with heated bed and chamber, whereas its firmware was customized from an open version of Marlin Kimbra.

The USB camera was installed with a field of view perpendicular to the build plate. This was done by removing the top cover of the printer and placing a support bar to suspend the camera in the middle of the print bed over the A4v3 frame (Fig. 1). This position was chosen not to modify the printer head to preserve normal operative conditions of the printer when the monitoring system is not used.

During installation, the focus of the camera was manually adjusted to the top of the last printed layer, which is right below the end of the extrusion nozzle because the print bed lowers layer after layer. To capture a clear image of an entire layer, it is necessary to move the print head away from above the part. This movement is exploited to home the extruders along the *X* and *Y* axes, and an end switch is placed close to the home position. The switch is connected to an Arduino UNO Board for taking a synchronized picture of the print bed using the USB camera.

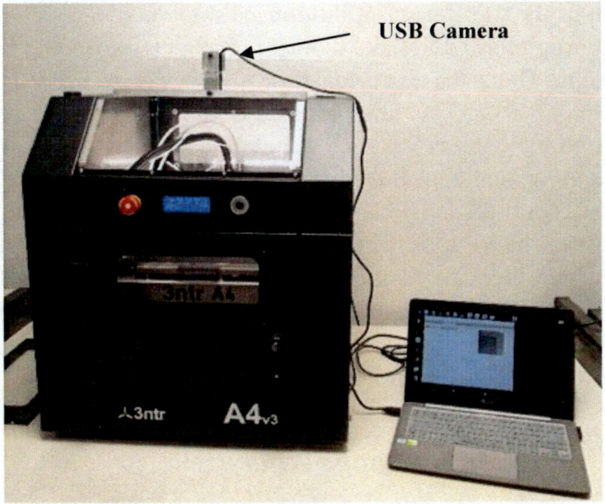

Fig. 1 Printer setup with the Svpro camera

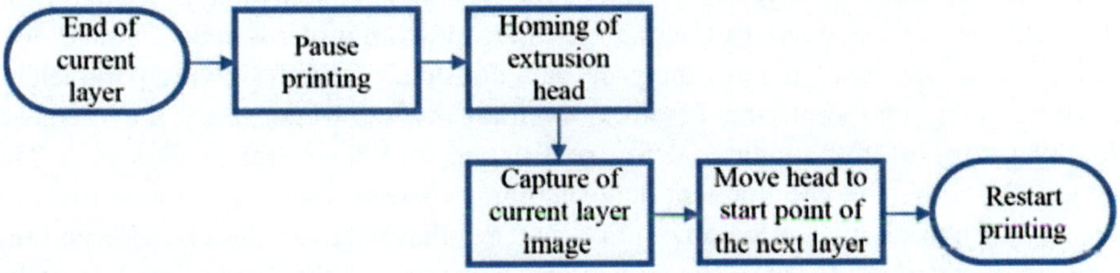

Fig. 2 In-situ monitoring procedure for each printed layer

The steps of the monitoring method for each layer are shown in Fig. 2. Before starting the 3D printing process, these steps are implemented by modification of the G-code file using a MATLAB script.

2.1 Modification of the Standard G-Code File

The 3D solid model of the part in the STL format is imported into the slicing software for the generation of the printing path. Kisslicer software is used for the A4v3 printer. Depending on the adopted layer thickness and printing parameters, the slicer creates the G-code file for a standard job without process monitoring. Then, the MATLAB script reads the text of the G-code file line by line, i.e., command after command. The script detects the end of each layer in the G-code when the z coordinate changes for the new layer because the z value is incremented by the layer thickness.

At this point in the file, the script introduces some new lines and commands for the layer monitoring phase. First, the filament is retracted from the extrusion nozzle to pause the print. Then, the extrusion head moved away from the print area by homing the machine axes. This idle motion is required to have a clear view of the current layer from the camera position. When the extruders are homed, they touch the additional end switch that sends the command to an external PC for taking the photo of the layer with the USB camera. Then, the print head is moved again over the printed layer to the location where the printing path for the next layer starts in the standard G-code file. The filament is fed again into the nozzle, and printing is resumed with the same lines and instructions of the standard G-code until the end of the next layer.

2.2 *Defect Detection by Image Analysis*

To detect the presence of printing defects within each layer, the corresponding layer image captured by the USB camera is compared to an ideal defect-free reference image using Matlab. The reference image for each layer is generated in Matlab from the print path of the standard G-code file. Starting from a blank image, a marker is used to trace every segment of the print path considering the following parameters: starting point, endpoint, the length E of filament to be extruded, the diameter D (2.85 mm) of the filament before extrusion, and the layer thickness t. The expected width W of the filament deposited along the segment is computed from a balance of mass between the amount of material that is fed to the extruder and the one that is extruded through the nozzle and deposited in the print area. The width W can be calculated by Eq. 1.

$$W = \pi(D/2)^2 \cdot E/(t \cdot d) \tag{1}$$

where d is the length of the segment or travel distance of the extruder and nozzle between the starting point and the final point. All the segments of the printing path are traced with the corresponding width to form the reference image that represents the ideal expected cross section of the part in the considered layer. Differences between the reference image and the captured image of each layer are calculated through pixel-by-pixel comparison and are identified as defects.

During printing, real-time information is provided to the user via a graphical interface (Fig. 3a) specifically designed to report any defects that are detected in the current layer. This application is designed using Matlab App Designer to show the layer information through images with tools for their analysis.

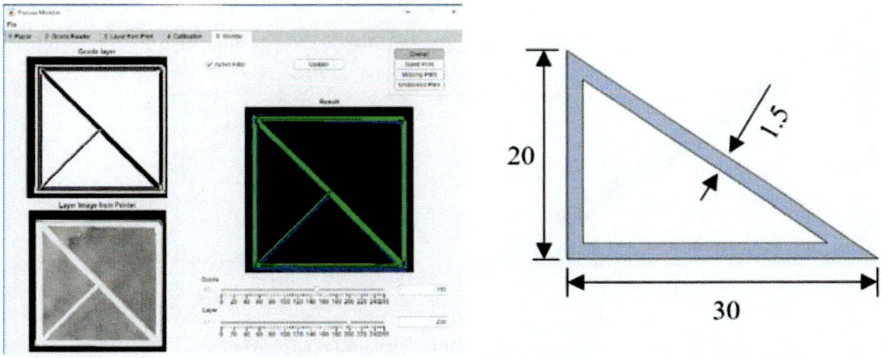

(a) (b)

Fig. 3 User interface of the monitoring system for defect detection (**a**); cross section of the test sample geometry with overall dimensions in millimeters (**b**)

3 Experimental Test

A simple triangular test sample (Fig. 3b) is printed with the scope of validating the in-situ monitoring system and the corresponding procedures. The two legs of the triangle measure 20 mm and 30 mm, respectively. The sample has a thickness of 1.5 mm, and its height is 0.5 mm in the build direction.

The sample was printed in ABS material using a yellow filament, while the printing parameters are summarized in Table 1.

The resolution of the USB camera was set to 800 pixels by 600 pixels over an area of 90 mm × 68 mm of the A4v3 printing bed. Therefore, 8.8 pixels of the camera image correspond to approximately 1 mm, so the monitoring system has an accuracy of about 0.114 mm in the detection of defects.

The defect-free reference image generated from the G-code file of the triangular sample is shown in Fig. 4a, whereas Fig. 4b shows the printing path without assignment of the extrusion width.

Before printing, the G-code file was modified on purpose to introduce an artificial defect by temporarily pausing the extrusion of the yellow ABS filament alongside a portion of the hypotenuse of the sample in all layers.

The photo that was captured by the in-situ monitoring system for the printed sample with the defect is shown in Fig. 5a. The corresponding results of the image

Table 1 Printing parameters for the test sample

Parameter	Value
Layer height	0.3 mm
Nozzle temperature	255 °C
Bed temperature	120 °C
Extrusion width	0.4 mm

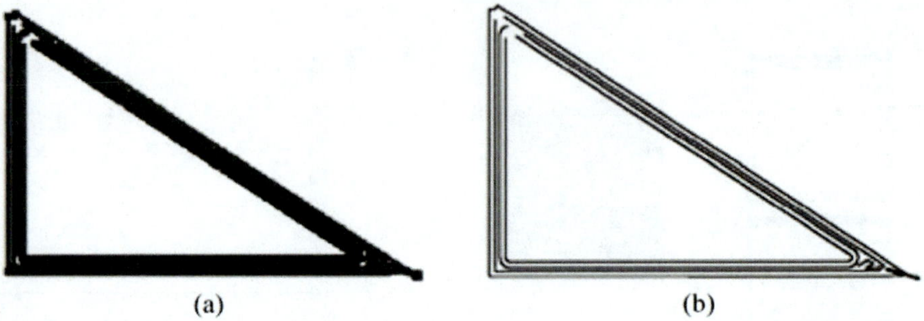

Fig. 4 Defect-free reference image for the triangular sample (a); printing path of the triangular sample without considering the extrusion width (b)

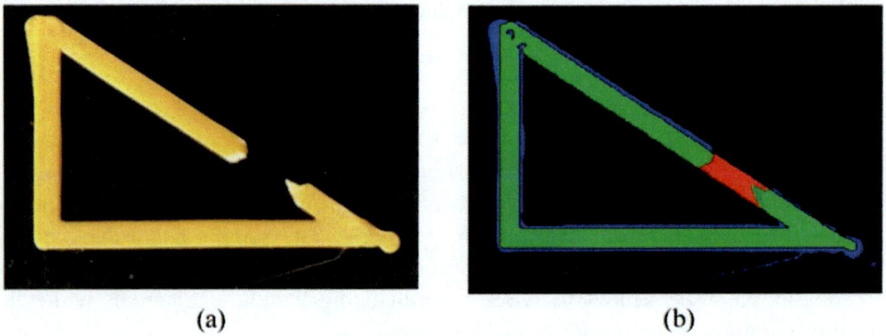

Fig. 5 Printed sample with artificial defect from paused extrusion alongside the hypotenuse (a); results of image analysis for detection of printing defects (b)

analysis for the detection of printing defects are reported in Fig. 5b. The green pixels in Fig. 5b represent the area where the material was deposited correctly as expected for the ideal defect-free part according to the reference image of Fig. 4a. The red pixels in Fig. 5b show the area of missing material because of the temporarily paused extrusion, whereas the blue pixels that appear along the borders of the green area are associated with an excess of material that was deposited across the edges of the sample. Further image processing allows isolating and separating these last two effects for detecting (i) missing material (Fig. 6a) and (ii) unwanted filament deposition (Fig. 6b).

For the most critical defect of missing material, the value of the area of the red pixels in Fig. 5b can be computed considering the resolution of the monitoring system. The defect that was artificially introduced by modification of the G-code covers 8 mm^2 of the surface corresponding to the hypotenuse of the triangular test sample (Fig. 6a).

The other type of defect is the excess of material deposited around the edges of the sample part. This defect is due to over extrusion of ABS material when the print

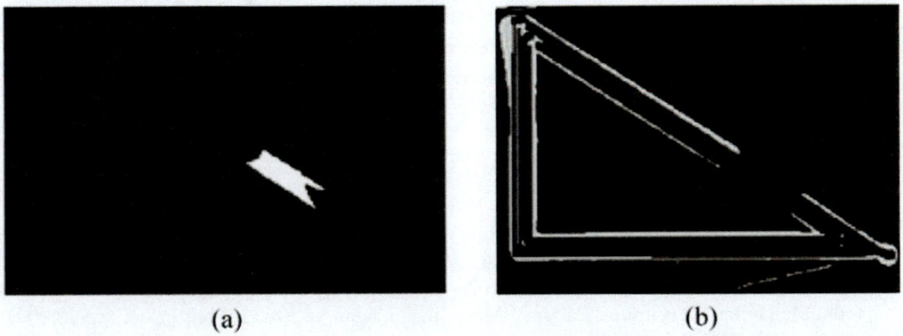

Fig. 6 Area of missing material corresponding to the artificial defect alongside the hypotenuse (a); excess of material deposited across the edges of the triangular test sample (b)

head slows down at the path points where a change in the direction of motion occurs. The over extrusion covers an area of about 48.5 mm^2 (Fig. 6b) wider than the nominal cross-sectional surface of 115.2 mm^2 of the CAD model of the test sample. This second defect was not artificially introduced through modification of the G-code, and it might be eliminated or reduced by tuning the printing parameters for better control of the material flow rate and deposition.

4 Conclusions

In this paper, a computer vision system was developed for in-situ monitoring of printing defects in open 3D printers. The monitoring system comprises a USB camera and an Arduino UNO board for automatic capturing of layer images of the deposited filament over the print bed with a cost of less than 100€.

The procedure requires access to the printing G-code file that must be modified for introducing the inspection step before printing each new layer. For the detection of the defect within every layer and cross section of the printed product, image analysis algorithms were implemented in Matlab software together with the development of a customized user interface. The presence of printing defects is detected by comparison of the camera image with an ideal defect-free reference image generated directly from the printing path of the G-code file.

The system was successfully tested on an A4v3 printer by 3ntr while printing a simple triangular geometry as a test sample. Defects related to both (i) missing material and (ii) undesired extrusion of excess material were identified with a resolution of 0.114 mm.

The in-situ monitoring system can be easily installed in other open 3D printers. In future works, the accuracy of the proposed system can be further improved by increasing the USB camera resolution to 2 Mpixels and the zoom setting. At a double cost, a second camera can be installed to exploit stereoscopic vision for getting 3D information about every deposited layer.

References

1. Rosen DW (2014) Research supporting principles for design for additive manufacturing. Virtual Phys Prototyp 9(4):225–232. https://doi.org/10.1080/17452759.2014.951530
2. Minetola P, Priarone PC, Ingarao G (2020) Sustainability for 3DP operations. In: Eyers D (eds) Managing 3D printing. Palgrave Macmillan, Cham, pp 97–126. https://doi.org/10.1007/978-3-030-23323-5_7
3. Everton SK, Hirsch M, Stravroulakis P, Leach RK, Clare AT (2016) Review of in-situ process monitoring and in-situ metrology for metal additive manufacturing. Mater Des 95:431–445. https://doi.org/10.1016/j.matdes.2016.01.099
4. Nuchitprasitchai S, Roggemann M, Pearce JM (2017) Factors effecting real-time optical monitoring of fused filament 3D printing. Prog Addit Manuf 2:133–149. https://doi.org/10.1007/s40964-017-0027-x
5. Grasso M, Colosimo BM (2017) Process defects and in situ monitoring methods in metal powder bed fusion: a review. Meas Sci Technol 28:044005. https://doi.org/10.1088/1361-6501/aa5c4f
6. Delli U, Chang S (2018) Automated process monitoring in 3D printing using supervised machine learning. Proc Manuf 26:865–870. https://doi.org/10.1016/j.promfg.2018.07.111
7. Paraskevoudis K, Karayannis P, Koumoulos EP (2020) Real-time 3d printing remote defect detection (stringing) with computer vision and artificial intelligence. Processes 8(11):1464. https://doi.org/10.3390/pr8111464
8. Petsiuk A, Pearce J (2020) Open source computer vision-based layer-wise 3D printing analysis. Addit Manuf 36:101473
9. Calignano F et al (2017) Overview on additive manufacturing technologies. Proc IEEE 105 (4):593–612. https://doi.org/10.1109/JPROC.2016.2625098
10. Bochmann L, Bayley C, Helu M, Transchel R, Wegener K, Dornfeld D (2015) Understanding error generation in fused deposition modeling. Surf Topogr Metrol Prop 3(1):14002. https://doi.org/10.1088/2051-672X/3/1/014002
11. Galati M, Minetola P (2020) On the measure of the aesthetic quality of 3D printed plastic parts. Int J Interact Des Manuf 14:381–392. https://doi.org/10.1007/s12008-019-00627-x
12. Galati M, Minetola P, Marchiandi G, Atzeni E, Calignano F, Salmi A, Iuliano L (2019) A methodology for evaluating the aesthetic quality of 3D printed parts. Proc CIRP 79:95–100. https://doi.org/10.1016/j.procir.2019.02.018

Machining and Welding

Highly Accelerated Life Test for High Speed Spindle Reliability

Lanzhi Liang, Weike Guo, Huawei Zhang, Hao Chen, Ruediger Heim, and Qun Lei

1 Introduction

Complex mechatronic system such as high speed spindle is typically used for high precision machining industry. Integrating the motor and spindle in one unit can reduce resource for couplings, belts and gearboxes. Spindle inertia can also be lowered and acceleration and deceleration times can be increased. To improve product performance in aspect of reliability, a life cycle test under real time condition is needed. However, it is not feasible to obtain the reliability through normal life cycle test since such test is neither time nor cost efficient. Therefore, an accelerated life test is introduced. The idea of the accelerated life test was first brought out by Yurkowsky et al. [1]. Previously, Zhu et al. [2] built a model of accelerated life test for a high speed grinding motorized spindle through strengthening loads; Hu et al. [3] indicated the influences of accelerated stress selection, maximum acceleration stress levels, failure criteria and small sample analysis in the reliability test; Jiang et al. [4] evaluated the spindle reliability using extremely small-scale sample based on false lime time distribution; Chen et al. [5] developed an accelerated test platform that can apply different load spectrums to the electric spindle under simulated working conditions; Lu et al. [6] presented a vibration

L. Liang · W. Guo (✉) · H. Zhang · H. Chen
Guangdong Institute of Intelligent Manufacturing GIIM, Guangdong, People's Republic of China
e-mail: guoweike@163.com

R. Heim
Fraunhofer Institute for Structure Durability and System Reliability LBF, Darmstadt, Germany

Q. Lei
Guangzhou Haozhi Industrial Co., Ltd., 511356 Guangzhou, People's Republic of China

R. K. Agarwal (ed.), *Recent Advances in Manufacturing Engineering and Processes*, Lecture Notes in Mechanical Engineering,
https://doi.org/10.1007/978-981-16-3934-0_8

noise signal-based analysis by selecting appropriate frequency band to monitor bearing conditions. Such idea can be integrated in the accelerated life test to determine the critical point of spindle failure; Williams et al. [7] used accelerometers and acoustic emission sensors for trending and diagnostic capability to describe the bearings behaviour through a life test. The study performed by Wang et al. [8] concluded that by increasing the interference between bearing inner ring and the shaft can improve bearing radial stiffness. To maximize, the bearing stiffness in both radial and axial directions requires careful considerations on the composition of interference and pre-tightening force.

The high speed spindle analysed here consists of different subsystems such as shafts, bearings, housings, cooling systems and motor, as shown in Fig. 1. The spindle has a maximum rotational speed of 60,000 RPM and was designed to serve 8000–10,000 h without breaking down. In reality, this specific type of spindle has a mean time between failures MTBF of less than 3000 h which is far below the requirement. In order to perform the accelerated life test for the bespoken spindle system, analysis on system and failure was firstly presented. From these analyses, the major failure mode was determined as bearing defect. Therefore, the work presented here used bearing life time to predict spindle life. Different design alternatives based on varies bearing parameters were then proposed to perform the accelerated life tests, and the results were compared to obtain the best design alternative for further product development.

Fig. 1 Spindle diagram

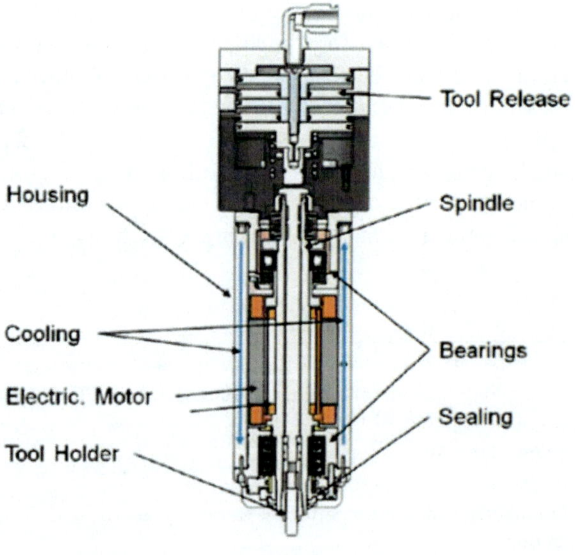

2 Spindle Reliability Assessment

2.1 *System Analysis*

The spindle consists of an inner and an outer shafts that are rotating simultaneously. The inner shaft is directly attached to the tool holder and a spring that counteracts the tool release mechanism. The outer shaft is suspended to the housing by three angular ball bearings with a minimum of 0.1 mm interference fit. The bearings have a work angle of 15°, which allows the bearings tolerate axial preload in order to achieve a suitable stiffness in axial direction. Maximum rotational speeds of 60,000 RPM could be achieved by the ceramic rolling elements inside the bearings under ideal conditions. The upper bearing is suspended by an array of preloaded disc springs to achieve a constant axial preload of 120 N. During the milling and drilling processes, electrical and mechanical energy from electric motor and bearing were dissipated into waste heat that needed to be cooled down to prevent inadequate temperature levels. High temperature could lead to damage to the relevant components. A close loop cooling system is integrated into the housing. A continuous flow of fresh air was led through the housing, exiting at the gap between the tool holder and the tip of the housing. Dust and cooling water penetrations to the tool must be prevented as contamination of coolant and dust would lead to corrosion and wear out. A minimum air flow rate of 30 l per minute was required in order to assure a satisfactory seal effect.

2.2 *Failure Analysis*

The frequency of occurrence and the root causes of different failure modes were determined using failure results and statistics.

Two sets of relevant failure data derived from returning spindles after failed were given. The first set as shown in Table 1 was the number of failures and sales during one-year period.

The second set of data indicated the frequency of occurrence of three distinct failure modes within a year as shown in Table 2.

Since most of the damaged tool holder were due to misuse by the customer, such as unwanted impact on the spindle tip during operation or incorrect setup. Roughly 92% of all failures recorded were due to bearing failures, leading to the assumption that the design of the bearing system may not be appropriate to sustain the

Table 1 Failure rate within a year

Failure rate (within 1 year)	
Number of sales	11,596
Number of failures	1920
Failure rate	16.56%

Table 2 Percentages of different types of failure

Type of failure	Units	Percentage
Upper bearing failure	1275	66.4
Lower bearing failure	493	25.7
Damaged tool holder	152	7.9
Total	1920	100

application loads. Improving bearing performance was the critical action to enhance the entire spindle system to fulfil design requirement.

Classic bearing failure modes are related to either material fatigue or wear. Material fatigue limits service life because of material defects under cyclic loading that creates spalls. Parameters for the specific type of bearing used in the spindle system were given in Table 3.

The bearing spacing, that is the distance between the upper bearing and bottom pair bearing set, is 106 mm. Hence, the ratio of the bearing spacing to the bore diameter *l*/*d* is larger than 6 which characterized this system to have a comparatively large bearing spacing. Due to cost and time efficiency and application constraints, the large bearing spacing of the specific spindle cannot be changed. Other parameters should be considered.

In order to check the fatigue strength, the dynamic load safety factor was calculated as:

$$S = C_0/P_0^* \tag{1}$$

where C_0 is the basic static load rating and P_0^* is the equivalent load from the dynamic load forces.

For the bearings used in the specified spindle which had a pair mounted DT together with a single bearing near the rear end (alignment of // \), the proportions of load on the most heavily loaded bearing were:

the axial load:

$$F_{0a} = 0.5 \cdot F_a \tag{2}$$

Table 3 Parameters of the bearing used in current design

Parameters	Values
d—bore diameter	17 mm
D—outer diameter	35 mm
B—width	10 mm
r—ball radius	0.3 mm
C_r—Basic dynamic radial load rating	8600 N
C_{0r}—Basic static radial load rating	2650 N
C_{ur}—Fatigue limit radial load	255 N
Bearing spacing	106 mm

and the radial load:

$$F_{0r} = 0.6 \cdot F_r \tag{3}$$

where F_a was 120N by the static pre-tightening force, F_r was 85 N maximum form grinding aluminium plates. Neglecting the comparatively small operational axial loads, with the pre-tightening force and the resultant radial forces, the load ratio F_{0a}/F_{0r} is certainly larger than 1.09 which means that for bearings with a contact angle of 15°, the equivalent load P_0^* was calculated as:

$$P_0^* = 0.5 \cdot F_{0r} + 0.46 \cdot F_{0a} = 57.7\,\mathrm{N} \tag{4}$$

With the C_0 was given as 2650 N by the bearing manufacturer, the dynamic load safety factor S was larger than 8, and the fatigue limit load in radial direction given in Table 3 was 255 N. These number demonstrated that the operating life is not restricted by metal fatigue in this application. Hence, the bearing damage modes had to be related to wear rather than material fatigue.

3 Accelerated Life Test Design

Potential cause of wear could be exceeding machine limits, leading to the consideration of three parameters: bearing types, pre-tightening force and interference between bearing inner ring and the shaft. The upper bearing was supported by a pre-tightening force of 120 N and the contact angle of 15°. Inner ring the bearing was mounted with a 1 μm interference to the shaft. Therefore, different alternatives are determined in Table 4.

Since there were various different parameters which might influence the operational bearing life, those effects had to be examined carefully, and economic constraints have to be respected too. To get results in a quick, cheap and repeatable way, an ex-centric mass which created a rotating force of 400 N at 50,000 rpm was designed and mounted on the spindle as shown in Fig. 2.

Performance criteria in such a test would be the temperature at the upper bearing (end of shaft), which was damaged more often than the pair mounted bottom bearings, as well as the vibration level at the nose of the spindle. These data were used for the evaluation of the health state. The test flow chart of the test is shown in Fig. 3.

Table 4 Design alternatives in bearings arrangement for the test

No.	Bearing type	Pre-tightening force (N)	Interference (μm)
1	Ball bearing with 15° angle	120	1
2	Ball bearing with 15° angle	70	1
3	Ball bearing with 25° angle	120	1
4	Ball bearing with 25° angle	70	3

Fig. 2 Ex-centric mass mounted on tool holder

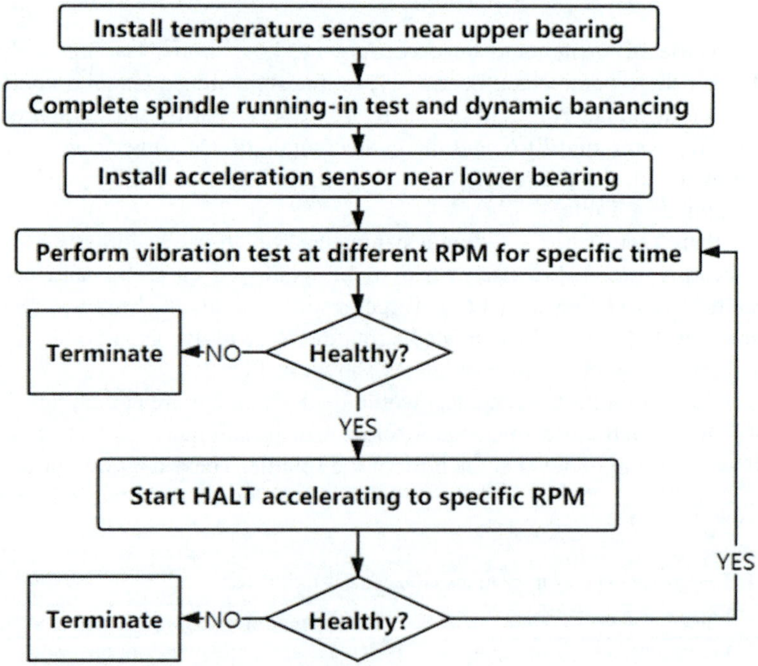

Fig. 3 Flow chart of the highly accelerated life test

Performance and reliability of the bearing system were evaluated by looking at the test duration at which temperature and vibration level indicate an issue regarding the health state of the bearing system. The overall vibration level was used by measuring the RMS acceleration data collected through transducer located as close to the baring as possible. The Crest Factor was computed as the ratio Peak/RMS. The development of a local fault led to peaks of the vibration signal but with little effect on the overall RMS level. As the fault progresses, more peaks will be generated causing the Crest Factor decrease and the RMS acceleration measure increase.

Data collection and computation on RMS and Crest Factor of acceleration should be processed after running-in and dynamic balancing test as the benchmark data for each alternative. The measurements would be trended over time and compared with the benchmark data.

4 Results and Analysis

The first design alternative failed after 160 h and the vibration levels for both upper bearing and tool sides of the design 1 spindle at initial and after 160 h are shown in Figs. 4 and 5, respectively.

From above, it was indicated that after a run time of 160 h the vibration level for the two bearing systems had increased quite significantly. The single upper bearing showed high acceleration throughout the complete frequency range, while for the tool side bearing system the acceleration level was increased more obviously. The acceleration level the upper bearing was in a worse shape after 160 h than the bottom bearings. For the bottom bearing system, the damage was much lower than that for the upper bearing which was identical to the reality.

Vibration levels after 160 h for each design alternatives were compared to the benchmark in order to find the best design. The vibration levels of the design 2 spindle at two positions at initial and after 160 h were shown in Figs. 6 and 7.

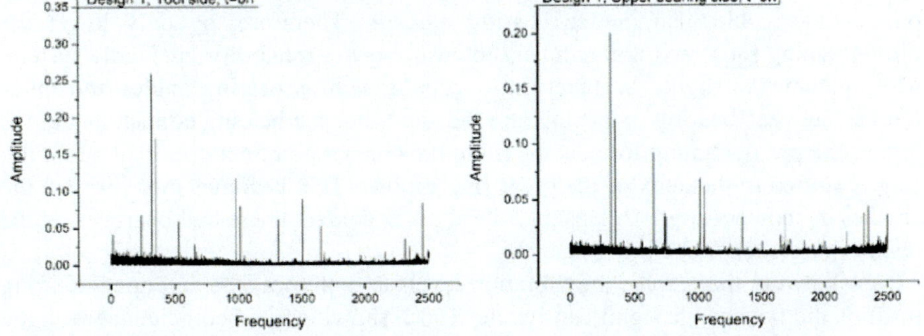

Fig. 4 Vibration level at $t = 0$ h for design 1 at tool side and upper bearing position

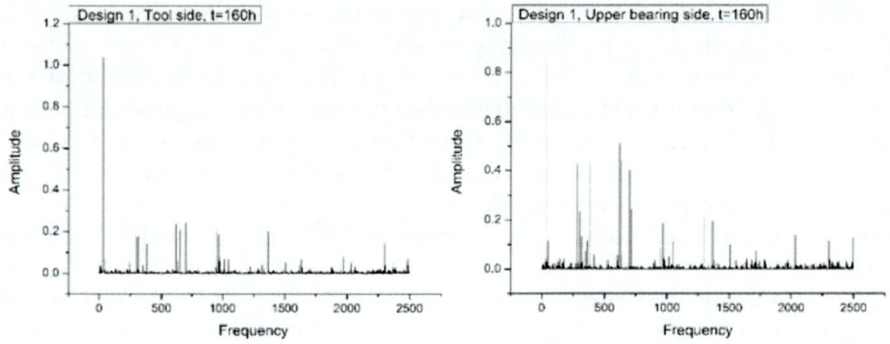

Fig. 5 Vibration level at t = 160 h for design 1 at tool side and upper bearing position

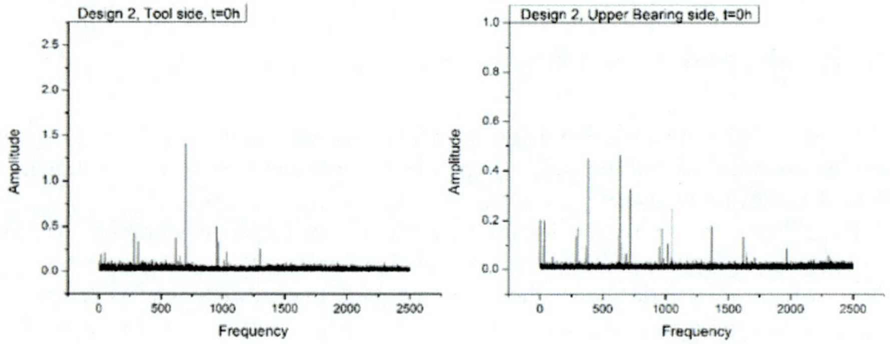

Fig. 6 Vibration level at t = 0 h for design 2 at tool side and upper bearing position

Lower pre-tightening force was introduced in design 2. It is clearly shown that the vibration level for design 2 at the beginning was larger than that for design 1. Reason for that is because the reduction in stiffness in axial direction. Pre-tightening force ensured the rigidity of the entire system as the initial vibration level for bottom bearings at tool side was higher than that in design 1. However, overstress on bearings could also induce bearing damage. Therefore, by only lower the pre-tightening force was not sufficient to ensure better reliability in spindle system. More parameters should be taken into account such as bearing angles and interference between bearing inner ring and the shaft. For the bearing contact angle, the higher the pre-tightening force is the more the contact point between ball and outer ring is shifted to the apex of the outer ring groove. That becomes more critical the smaller the contact angle is. Design 3 used a 25° contact angle ball bearing and the results were shown in Figs. 8 and 9.

As shown in the graphs, the vibration levels at both tool side and upper bearing position did not change significantly after 160 h showing good coincidence with the expectation. Since interference is a key parameter to ensure system rigidity, small

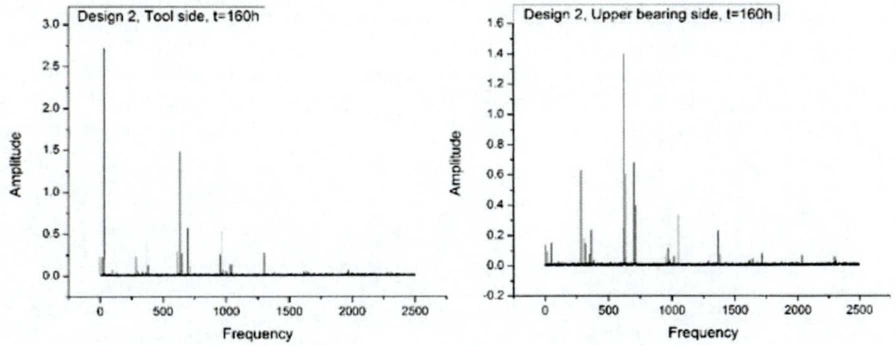

Fig. 7 Vibration level at t = 160 h for design 2 at tool side and upper bearing position

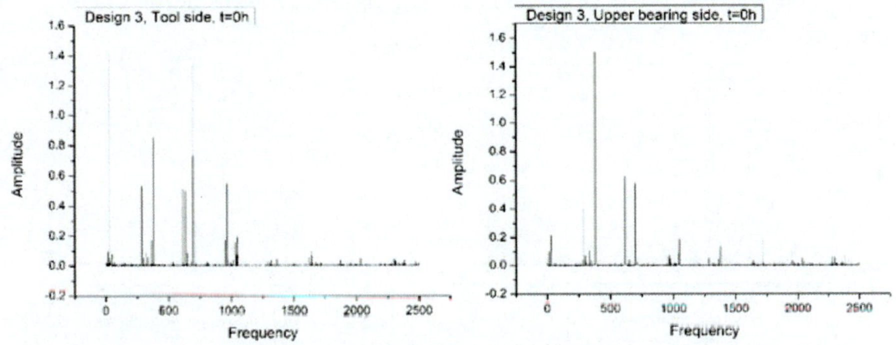

Fig. 8 Vibration level at t = 0 h for design 3 at tool side and upper bearing position

interference requires large pre-tightening force. Therefore, lowering pre-tightening force should be applied together with increasing interference. Design 4 used the 25° contact angle ball bearing the same as design 3 with lowered pre-tightening force and enlarged interference. Results on vibration levels at two positions at initial and after 160 h were shown in Figs. 10 and 11, respectively.

Vibration level at upper bearing position was decreased remarkably showing good improvement. However, an increase in vibration level at tool side indicated a poor performance of bottom bearings. Since the 25° contact ball bearing was only applied at upper bearing position, bottom bearings had to compromise the lowered pre-tightening force. Therefore, in further analysis, the composition of bottom bearings, lowered pre-tightening force and the interference between the bearing inner ring and the shaft should be considered. Comparison of all design alternatives showed that by changing the bearing contact angle from 15° to 25°, lowering the pre-tightening force and increase the interference between the bearing inner ring and the shaft had a significant impact on spindle performance.

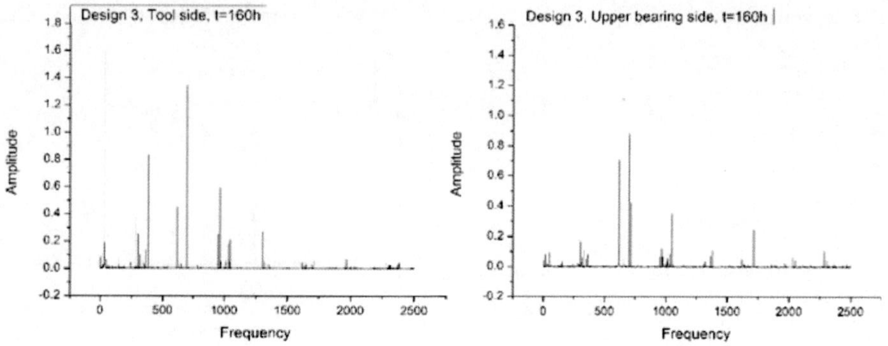

Fig. 9 Vibration level at t = 160 h for design 3 at tool side and upper bearing position

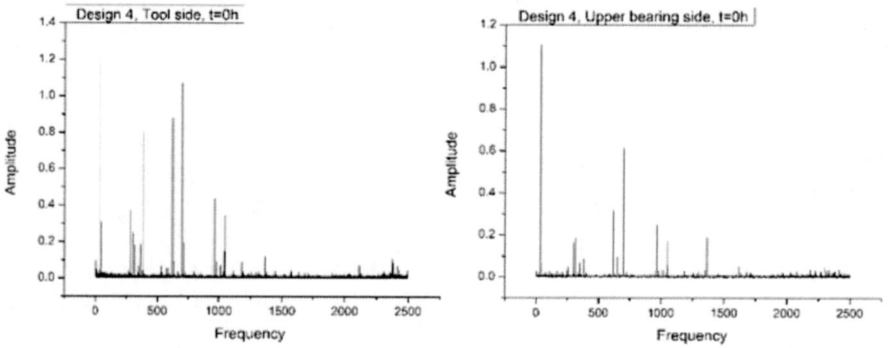

Fig. 10 Vibration level at t = 0 h for design 4 at tool side and upper bearing position

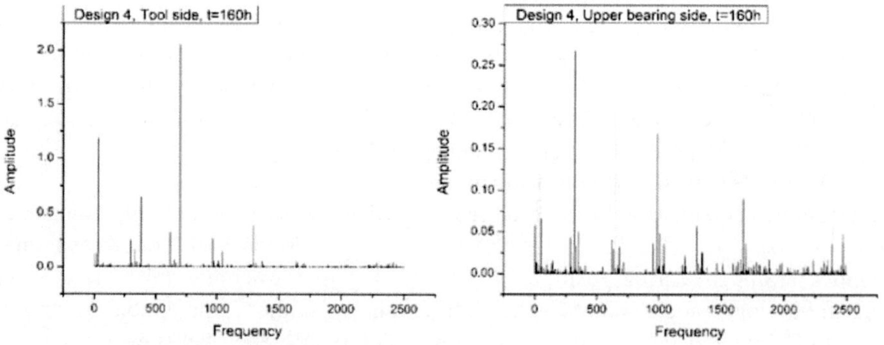

Fig. 11 Vibration level at t = 160 h for design 4 at tool side and upper bearing position

5 Conclusion

A highly accelerated life test on a high speed spindle system was carried out in order to improve its reliability. Based on system and failure analysis, parameters used on upper bearing were found to be the major failure mode of spindle failure. Hence, four design alternatives based on varies different bearing parameters were designed for the test. Results can be concluded that:

- Using ex-centric mass, which could enlarge radial load, as a tool used in highly accelerated life test for spindle is in accordance with the reality. Such test gives a good foundation to future tests which should run the spindle through failure;
- Reducing pre-tightening force is sufficient to improve vibration behaviour;
- Using 25° contact angle ball bearing instead of 15° contact angle improved the endurance to high pre-tightening force;
- Composition of 25° contact angle ball bearing, lowered pre-tightening force and the enlarged interference between the bearing inner ring and the shaft has increased vibration behaviour at upper bearing position;
- Bottom bearings and their compromise with lowered pre-tightening force and the interference should be considered in future analysis;
- In this paper, the data of the 4 design alternatives were from the beginning to 160 h. In future experiments, testing of the spindle through failure should be implemented to obtain full failure data.

Acknowledgements This work was supported by Science and Technology Planning Projects of Guangdong Province (No. 2018A050506057), Science and Technology Planning Projects by Guangzhou (No. 201804010170 and No. 201807010023) and the research fund of special projects by Guangdong Academy of Science (No. 2019GDASYL-0105070 and 2019GDASYL-0105072).

References

1. Yurkowsky W, Schafer R, Finkelstein J (1967) Accelerated testing technology handbook of accelerated life testing methods, vol 1. Gordon Associates INC., New York
2. Zhu D, Liu H, Yuan D, Liu L (2013) Accelerated reliability life testing analysis of high-speed grinding motorized spindle. J Mech Strength 35:493–497
3. Hu W, Zeng K, He X, Cui G, Yuan S, Liu E (2015) Review of reliability test for machine tool spindle. Mod Manuf Eng 06:80–83
4. Jiang X, Liu H, Liu L, Zi J (2013) Extremely small-scale sample's reliability of an electric spindle based on distribution of false lifetime. J Vib Shock 32:80–85
5. Chen C, Chen F, Ye Y, Zhang H, Yang Z, and Du X (2016) Design of accelerated reliability test for CNC motorized spindle based on vibration signal. In: 3rd international conference on industrial engineering and application. MATEC Web Conf. 68(2016)02001
6. Lu C, Wang G, Yie H (2003) Gap conditions monitoring for main beairng based on noise measurements. J Vib Shock 39:33–36

7. Williams T, Ribadeneira X, Billington S, Kurfess T (2001) Rolling element bearing diagnostics in run-to-failure lifetime. Test Mech Syst Signal Process 15:979–993
8. Wang S, Xia Y (2006) Effect of the interference fit and axial preload in the stiffness of the high speed angular contact ball bearing. J Univ Sci Tech China 06:1315–1320

Predictions of Intermetallic Compounds in AC Pulse MIG of Dissimilar Materials Between Aluminium Alloy and Galvanized Steel by Numerical Analysis

Hee-Seon Bang, Yun-Hee Jo, and Hye-Seul Yoon

1 Introduction

Due to environmental changes in the automobile industry, it is necessary to improve fuel efficiency according to the CO_2 emission regulations. A solution is to use lightweight materials [1]. Numerous problems lie with welding of dissimilar materials, and not only is it difficult to weld because of the different chemical and physical properties but also because of the weakly created intermetallic compounds (IMCs) on the interface of the weld. Therefore, a low-heat input process is required to minimize IMCs, where AC pulse MIG is one of the low-heat input processes and is enable to minimize IMCs formation. Y. Kim et al. studied the dissimilar welding between steel and aluminium alloy using arc heat source. They reported that the most basic reason for the difficulty of fusion welding in Al alloy and steel is the formation of brittle IMCs [2]. Zhang et al. compared the growth of IMCs of tensile-shear strength of Al alloy and galvanized steel plates welded under low- and high-heat input conditions [3]. These experimental studies are actively conducted like the studies done by Marcin Kubiak and A. Das, etc. through numerical analysis. However, little detailed research has been conducted on the relationship between the EN ratio, i.e. the heat input and the IMCs [4, 5]. Therefore, this study aims to investigate the thickness of the layers of IMCs according to the current, i.e. the heat input by applying metal inert gas (MIG) welding for Al alloy (AA6061-T6) and galvanized steel (GI steel) through numerical analysis.

H.-S. Bang
Department of Welding and Joining Science Engineering, Chosun University, Gwangju 61452, Republic of Korea

Y.-H. Jo · H.-S. Yoon (✉)
Department of Welding and Joining Science Engineering, Graduate School, Chosun University, Gwangju 61452, Republic of Korea
e-mail: hyeseul@chosun.kr

R. K. Agarwal (ed.), *Recent Advances in Manufacturing Engineering and Processes*, Lecture Notes in Mechanical Engineering,
https://doi.org/10.1007/978-981-16-3934-0_9

2 Experimental Procedures

2.1 Numerical Model and Condition

In the presented study, the shape and the principal dimensions of the work piece are 200 mm (L) × 100 mm (W) × 2 mm (t) (AA6061-T6 and GI steel), which are the same in the numerical models. The experiment schematic and the dimensions of finite element model (FEM) are shown in Fig. 1. For numerical analysis, commercial software Ansys is used. The initial conditions were set as room temperature 300 K. Used welding conditions are 50, 60, 70 A of current at a welding speed of 8.33 mm/s. The analysis is conducted by a 3-dimensional heat transfer model, adopted to conduct heat conduction for the welded joints. Figure 2 shows the considering temperature dependence of the material properties. And the effect of moving the heat source according to the transfer speed of the welding arc was considered. The joint is meshed with a fine mesh of the solid elements, where the mesh size is 2 mm. Mesh type was modelling. The welding heat source moves within the time to consider the moving heat source. The heat input from the heat source is shown in Eq. (1) [6–8].

$$q = \frac{\eta EI}{v} \tag{1}$$

q is the heat flux (w/mm^3), η is the welding efficiency, E is the voltage (V), I is the current (A) and v is the volume of welding applied (mm^3). The thickness of the IMCs layer is estimated by extracting it from the temperature history of the welding interface and by comparing it with the actual measurement using a thermocouple so that the validity of the numerical analysis is verified.

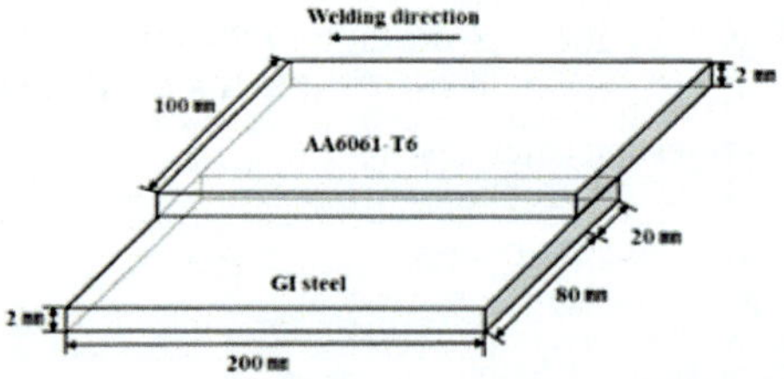

Schematic of AA6061-T6 and GI steel welded joints specimen

(a) Analysis model

(b) Front view

Fig. 1 Schematic of welded joints specimen and configuration and mesh division of FE analysis model

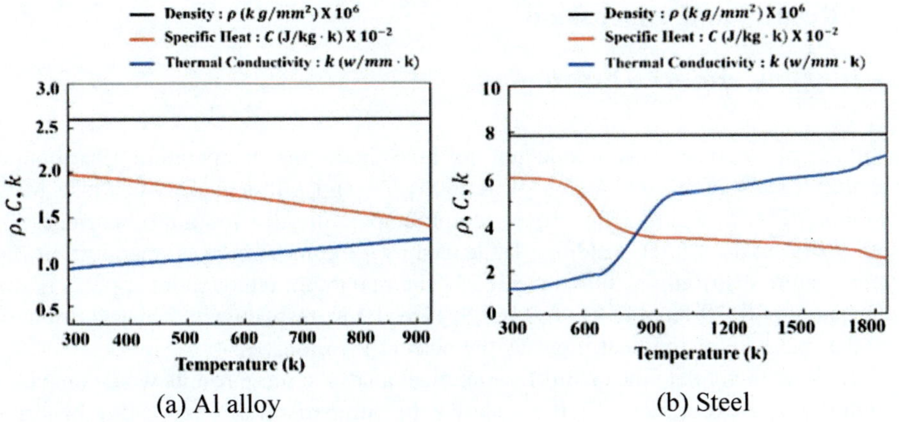

Fig. 2 Temperature dependence of material properties

2.2 *Estimation of IMCs Layer Thickness*

When dissimilar materials are jointed, the IMCs are formed. The IMCs change their mechanical properties in this process. The IMCs of Fe–Al cause brittleness of the weld. Hence, the thickness of the IMCs layer is predicted through a numerical analysis. The growth of the IMCs layer is calculated with the parabolic law of diffusion in regards to the cumulative the interface temperature history [9].

$$\lambda = \sqrt{kt} = \sqrt{k_0 \exp(-\frac{q}{RT})t} \tag{2}$$

$$\lambda_n^2 = \lambda_{n-1}^2 + k_0 \cdot \exp\left[-\frac{q}{R\left(\frac{T_0+T_1}{2}\right)}\right] \cdot (t_n - t_{n-1}) \tag{3}$$

$$\lambda_n = \sqrt{\lambda_{n-1}^2 + k \cdot \exp\left[-\frac{q}{R\left(\frac{T_n+T_{n-1}}{2}\right)}\right] \cdot (t_n - t_{n-1})} \tag{4}$$

λ is the IMCs layer growth, k_0 is the pre-exponential factor (1.32×10^2 m^2/s^{-1}), q is the activation energy (190–240 kJ/mol), R is the gas constant (8.341 J/molk) for the IMCs layer growth, T is the temperature and t is the corresponding diffusion time. IMCs grow according to Eq. (4). Temperature is measured using a thermocouple of K-type.

3 Results and Discussion

3.1 Temperature Distribution

Numerical analysis was conducted to investigate the temperature distribution characteristics of welded joints of AA6061-T6 and GI steel by AC Pulse MIG welding. Figure 3 shows the cross-section temperature distribution by current (50, 60, 70 A) at 12 s after welding. Table 1 shows the maximum temperature in the temperature distribution shown in Fig. 3. The maximum temperature appears in the current 50, 60, 70 A as 1043.96, 1167.9, 1259.19 K, respectively. This is attributed to the interface of the heat input as the current increases.

In order to validate the results of numerical analysis, these results were compared with those of experiment. Figure 4 shows the comparison of temperature histories between predicted values by numerical analysis and measured values for various currents. From the results, it is found that the numerical analysis and measured values are in agreement for temperature history's characteristics, and the measured values are slightly lower than the numerical analysis, which shows the almost identical characteristics, qualitatively and quantitatively.

3.2 Intermetallic Compounds Prediction

For the validity of the numerical analysis values, Fig. 4 shows the graph of the measured temperature by a thermocouple and the analysed temperature. It can be seen that the maximum temperature of numerical analysis of 50, 60, 70 A were 772, 848, 916 K. When the measured temperature and the analysis temperature are compared, there is a slight difference in the maximum temperature. Figure 5 shows the thickness of the IMCs layer according to the current observed by scanning

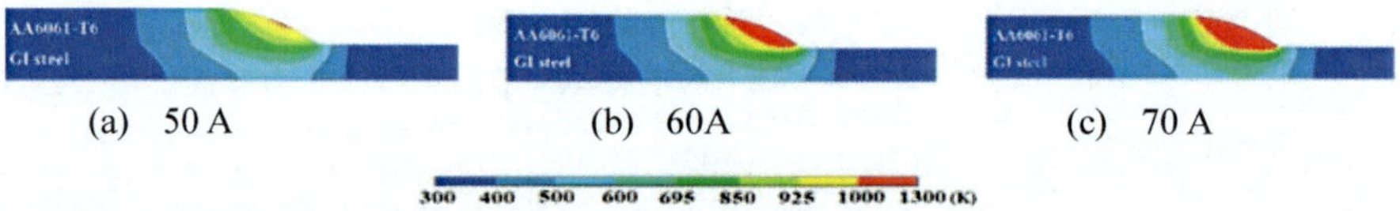

Fig. 3 Temperature distribution of cross section of welded joints for various currents

Table 1 Maximum temperature of cross section of welded joints for various currents

Current (A)	Maximum temperature (K)
50	1043.96
60	1167.9
70	1259.19

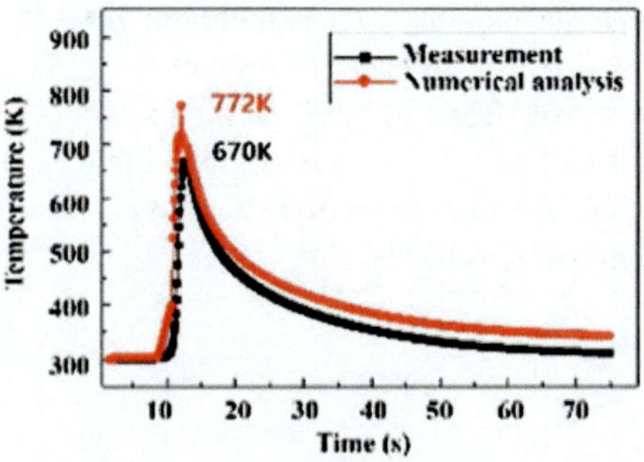

(a) 50 A

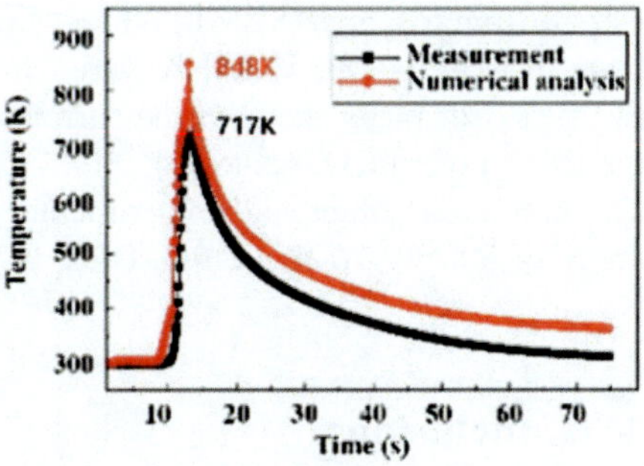

(b) 60 A

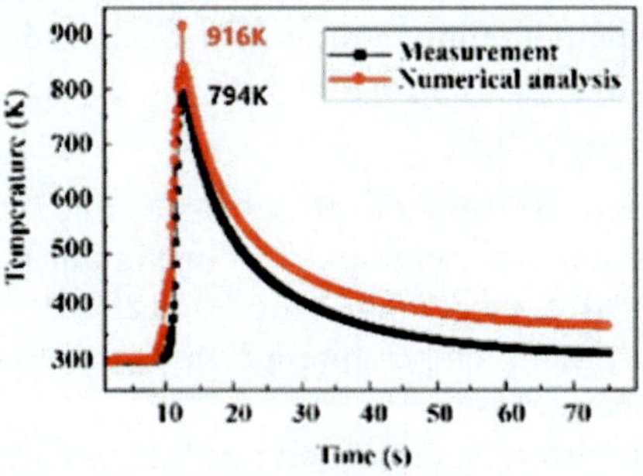

(c) 70 A

Fig. 4 Temperature histories of cross section of welded joints for various currents

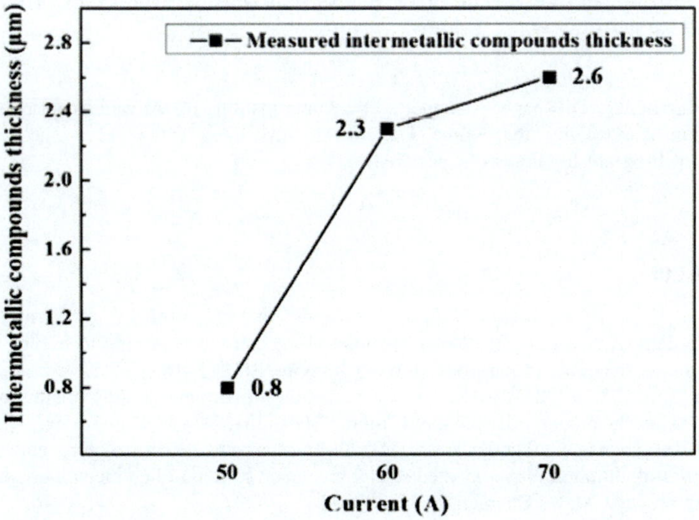

Fig. 5 Thickness of IMCs layer of welded joints for various currents

electron microscope (SEM). When the currents increased, the thickness of IMCs layer increased—50, 60, 70 A were 0.8, 2.3, 2.6 μm. In order to predict the growth of IMCs, Eq. (4) is used. As the currents increases from 50 to 70 A, the thickness of the IMCs layer increases to 1.4, 2.08, 3.39 μm due to the increase in heat input [10]. The predicted values and experimental values on thickness of IMCs layer, differences of 50, 60, 70 A are 0.6, 0.22, 0.79 μm, respectively.

4 Conclusions

In this study, the thickness of the IMCs layer causing brittleness in the welded joints of dissimilar materials (aluminium alloy and galvanized steel) is predicted using a numerical analysis and the comparison with the actual measurement. The results can be summarized as follows:

(1) At the temperature distribution of the same time, the maximum temperature increases as the current increases. The maximum temperature appears in the current 50, 60, 70 A as 1043.96, 1167.9, 1259.19 K, respectively. It can be inferred that maximum temperature increases due to the increase in heat input from the currents increase.
(2) The temperature history examination result for each current is 50, 60, 70 A as 772, 848, 916 K, respectively. The measured temperature is approximately 100–130 K different from the numerical analysis temperature.
(3) It is confirmed that the thickness of IMCs layer becomes thicker as the current increases. When comparing the thickness of IMCs layer with prediction and measurement, the differences is at each current is 50, 60, 70 A as 0.6, 0.22, 0.79 μm, respectively. The error is less than 1 μm, which can prove the reliability of the numerical analysis results.

Acknowledgements This paper is supported by Korea Institute for Advancement of Technology (KIAT) grant funded by the Korea Government (MOTIE) (P0008425, The Competency Development Program for Industry Specialist)

References

1. Kim Y, Park KY, Kwak SB (2015) Mechanical fastening and joining technologies to using multi mixed materials of car body. J Weld Joining 33(3):12–18
2. Kim Y, Kim CH, Lee BY (2018) A study on the dissimilar metal joining of aluminum to steel using the arc heat source(III). J Weld Joining 36(4):23–35
3. Zhang HT, Feng JC, He P, Hackl H (2007) Interfacial microstructure and mechanical properties of aluminium–zinc-coated steel joints made by a modified metal inert gas welding–brazing process. Mater Charact 58(7):588–592

4. Kubiak M, Piekarska W, Saternus Z, Domański T (2016) Numerical prediction of fusion zone and heat affected zone in hybrid Yb:YAG laser + GMAW welding process with experimental verification. Proc Eng 136:88–94
5. Das A, Shome M, Goecke S-F, De A (2016) Numerical modelling of gas metal arc joining of aluminium alloy and galvanised steels in lap joint configuration. Sci Technol Weld Joining 21 (4):303–309
6. Crucifix S, van der Rest C, Jimenez-Mena N, Jacques PJ, Simar A (2015) Modelling thermal cycles and intermetallic growth during friction melt bonding of ULC steel to aluminium alloy 2024-T3. Sci Technol Weld Joining 20(4):319–324
7. Kajihara M (2006) Quantitative evaluation of interdiffusion in Fe2Al5 during reactive diffusion in the binary Fe–Al system. Mater Trans 47(6):1480–1484
8. Bang HS, Oh CI, Ro CS, Park CS, Bang HS (2007) Analysis of thermal and welding residual stress for hybrid welded joint by finite element method. J Weld Joining 25(6):11–16
9. Crucifix S, van der Rest C, Jimenez-Mena N, Jacques PJ, Simar A (2015) Modelling thermal cycles and intermetallic growth during friction melt bonding of ULC steel to aluminium alloy 2024-T3. Sci Technol Weld Join 20(4):319–324
10. Cho SM, Kong HS (2003) The arc brazing by variable polarity AC pulse MIG welding machine. Korean Weld Joining Soc 21(4):56–62

Elucidate Fluid Vortex in Plasma Arc Welding

Thanh-Hai Nguyen, Nguyen Van Anh, Shinichi Tashiro, Thu Le Quy, and Manabu Tanaka

1 Introduction

Plasma arc welding is one advanced joining technology with superior characteristics such as: high current density, constricted and compressed arc plasma, and high plasma jet velocity. This is due to a fact that there is a copper orifice with a small diameter hole inside the torch construction. This orifice is with two functions: discharge the main arc plasma and orientate the plasma column in ring form and straight. This is an important fracture that is different to traditional TIG welding processes. In the past decades, PAW has had applications widely in manufacturing technology from steel structure [1], shipbuilding [2], automobile [3], airplane [4], to space shuttle [5], etc. Recently, it has received the enormous attention of scientists and engineers. In the 1970s, the National Aeronautics and Space Administration (NASA) of the US applied this technology to produce the rocket body [6]. In another effort, Boeing and GE companies have fabricated the aluminum body of the

T.-H. Nguyen
Faculty of Mechanical Engineering, Ho Chi Minh City University of Technology (HCMUT), 268 Ly Thuong Kiet Street, District 10, Ho Chi Minh City 700000, Vietnam

T.-H. Nguyen
Viet Nam National University Ho Chi Minh City, Linh Trung Ward, Thu Duc District, Ho Chi Minh City 700000, Vietnam

N. Van Anh (✉) · S. Tashiro (✉) · M. Tanaka
Joining and Welding Research Institute (JWRI), Osaka University, Osaka, Japan
e-mail: tashiro@jwri.osaka-u.ac.jp

N. Van Anh
Murata Welding Laboratory, Osaka 532-0012, Japan

N. Van Anh
Hanoi University of Science and Technology, Hanoi, Vietnam

T. Le Quy
National Research Institute of Mechanical Engineering, Hanoi, Vietnam

R. K. Agarwal (ed.), *Recent Advances in Manufacturing Engineering and Processes*, Lecture Notes in Mechanical Engineering,
https://doi.org/10.1007/978-981-16-3934-0_10

airplane by VPPA process [7]. In the 2000s, Zhang and his colleagues published several papers to study the plasma flux on the top surface and the deformation of keyhole channel [8]. These results have originated to apprehend the nature of this technology. On the other hand, it was indicated in many publications that thermodynamic fluid flow of melted domain is a very useful information to understand the mechanism in welding, additive manufacturing, and other material processes [9, 10]. In an effort to understand the thermodynamic phenomena of fluid flow in this welding process, Wu and his team have performed a lot of experiments and simulation with the support of high-speed imaging cameras (HSVC). The results have exploded that the keyhole exit information at the bottom side is great necessary to consider the stability of the melted domain [11, 12]. In another paper, Nguyen et al. have observed the behavior of the keyhole on the bottom surface. The results showed that during the arc plasma switch-off period, the keyhole contour is changed much and then it became stable at several milliseconds [13]. By diversification of shielding gas composition, Nguyen et al. have demonstrated that the physical nature changes of the arc plasma column caused the variation of the arc pressure and current density. This leads to variation of keyhole contour and effects on the diversion of fluid vortexes on the melted domain surface. And then, the Marangoni force was reversed [14, 15]. In another paper, the correlation among eddies in the weld pool is the main factor to cause the welding defects on the surface of jointed structure such as undercut [16]. In order to improve capability to join thick plates, a novel hybrid welding technology with the combination of PAW torch and MIG torch was tried. The results implied that the plate of 12 mm thickness can be welded one pass without root groove and gap between workpieces [17]. On the other hand, the study on the formation mechanism of physical metallurgy, microstructure behavior, and mechanical properties is an indispensable approach in studying the nature of material processes [18]. In the PAW process, this trouble has been studied insensitively by researchers. In a recent paper, the microstructure and mechanical properties of dual phase steel DP600 with thickness of 6 mm were discussed [19]. In another publication, the dissimilar material joint between ultra-high strength steel and aluminum alloy (Al–Si) was studied [20]. In an effort to apply the PAW process to weld aluminum and its alloy, microstructure, and properties of Al–Mg alloy were investigated with a variable polarity plasma arc (VPPA) variant [21]. Recently, in endeavors to drive the VPPA process in welding aluminum and its alloy, Bin et al. have tried to visualize molten pool images of A5052P aluminum alloy using tracer particles. The results have provided the essential knowledge to explode the sufficient identification about heat transport and convection in the melted domain [22].

Summary, there are a numerous publications of PAW process but no any publication considering the collection among the input parameters (welding parameters) with the arc plasma nature, the change of heat mass transfer, fluid flow thermodynamic nature, crystallization in weldment, microstructure behavior, and the mechanical properties has been presented well to understand insightfully about this technology. As a result, the production effectiveness of this technology is still quite low, and its applicable expansion was slow down a little bit. In order to elucidate complexity nature and increase the application scope of this technology, we have started to observe the plasma arc and the material flow of the weld pool [16]. The data have implied that the thermodynamic phenomenon is based on the heat input under the change of welding parameters [23]. Furthermore, the eddies inside the weld pool varied the profile and strength when the welding parameters were changed [24]. In addition, the relationship between the eddies and welding surface defects was considered with the change of welding current [25]. The copper orifice is the most important component in a PAW torch because this orifice can affect arc plasma characteristics. It is the fountainhead to cause the molten pool behavior, weldment crystallization, and microstructure. This is a main mechanism of welding processes. However, these phenomena have been not yet indicated in publications recently. In order to explode a new understanding of the relation between the fluid vortex in melted domain and the orifice diameter, we discuss, for the first time, the change of fluid vortex inside the melted domain under the change of copper orifice diameter in this paper.

2 Experimental Method

In this investigation, a plasma welding torch (100 WH, Nippon Steel Welding & Engineering Co., Ltd) in combination to a welding source (NW-300ASR, Nippon Steel Welding & Engineering Co., Ltd) were the main welding equipment. For protecting the backside of the weld pool, a back-shielding gas protector is attached on the backside of the workpiece. The shielding gas is supplied to this apparatus in order to protect the weld pool and the crystallization process without the negative influence of the air. The copper orifice diameter inside welding torch is 2.0 and 2.4 mm. Other welding conditions can be seen in Table 1.

In order to measure the fluid vortex of the melted domain, two X-ray observation systems using X-ray sources will transmit throughout the melted zone of the workpiece. And then, X-ray signals are printed on films (X-ray films). Afterward, these films are captured by two high speed imaging cameras, simultaneously. As a

Table 1 Welding parameters

Parameters	Value (unit)
Set back of tungsten electrode	3 (mm)
Welding DC current	120 (A)
Arc length	5 (mm)
Plasma gas flow rate	1.7 (l/min)
Shielding gas flow rate	3 (l/min)
Back-shielding gas flow rate	7.5 (l/min)
Welding velocity	3 (mm/s)
Workpiece	SUS304 thickness 4 (mm)

Table 2 Setup parameters of four-dimension observation system

Parameters	Value (unit)
X-ray source 1	225 (kV) 3.5 (mA)
X-ray source 1	230 (kV) 1.0 (mA)
HSVCs frame	1000 (pfs)
X-ray inclination angle	30 (degrees)
Tracer particle (tungsten) diameter	0.3 (mm)

result, four-dimension images of melted domains are detected. Details of setup parameters are found out in Table 2.

Afterward, four-dimension images are converted to real values of velocity, distance, dimension by tracer analysis software. In this case, two HSVC are controlled synchronously. A schematic diagram of this observation system can be seen in Fig. 1.

3 Experimental Results

3.1 Weld Bead Cross-Section and Keyhole Contour

Weld bead cross-section and keyhole contour are presented in Fig. 2a, b with copper orifice 2.0 and 2.4 mm, respectively. The width of the top bead is about 5.5 mm with nozzle 2.0 mm and about 6.9 mm with nozzle 2.4 mm. The width of the bottom bead is about 2.6 mm (2.0 mm orifice) and 2.9 mm (2.4 mm orifice). The keyhole diameter at the upper side of the weld pool is about 4.6 mm with nozzle 2.0 mm and 4.9 mm with nozzle 2.4 mm. On the other hand, the bead width is about 2.2 and 2.4 mm, respectively, at the lower side of the weld pool. The inclination angle of the rear kerf of the weld pool is largely increased in case of

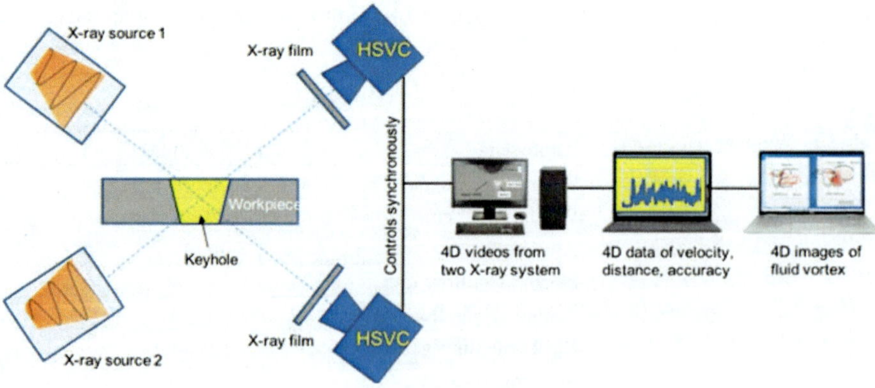

Fig. 1 Diagram of 4D X-ray observation and analysis system

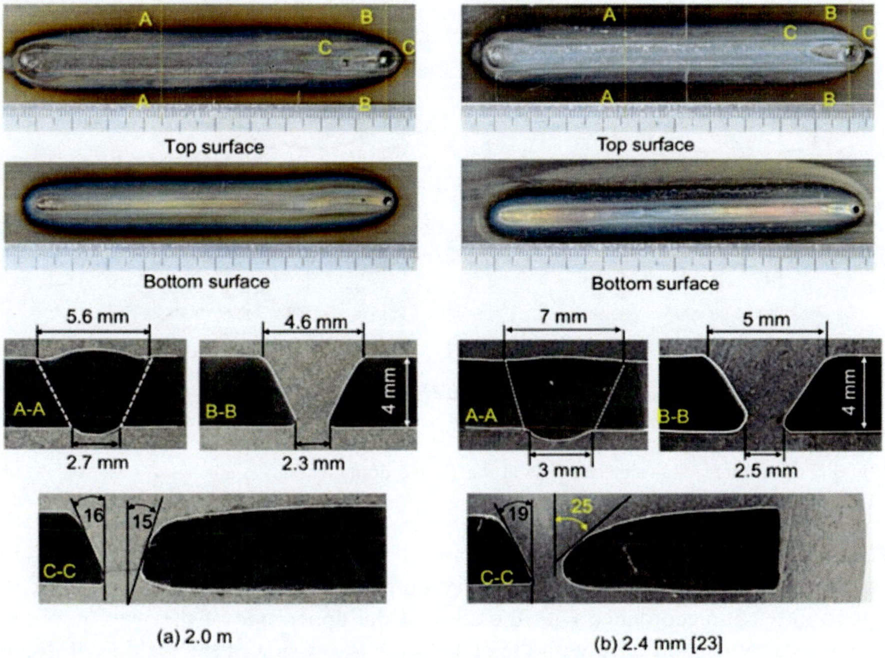

Fig. 2 Cross-section and keyhole contour

2.4 mm (25°), and meanwhile, it is about 15° with nozzle 2.0 mm. The inclination angle of the front kerf of the weld pool is about 16° in case of 2.0 mm and 19° with nozzle 2.4 mm. The deviation between the torch centerline and the keyhole centerline on the lower side of the weld pool was higher with nozzle 2.4 mm due to the deduction of plasma pressure and heat flux. Therefore, the keyhole and the weld bead were wider especially on the upper side in case of 2.4 mm. Furthermore, the bead appearance is well without welding defects such as undercut, convex, reinforcement, and burn through in both cases.

3.2 *Fluid Vortex Behavior in Melted Domain*

Four-dimensional vortex images of the melted domain in both cases of 2.0 and 2.4 mm are expressed in Fig. 3a, b, respectively. There are two movement directions of tracers at the melted domain in both 2.0 and 2.4 mm. After moving over the front part of the keyhole, some tracers motived in clockwise direction and backward at the upper side and some of them went in counter-clockwise and backward at the lower side of the weld pool. From this trend of tungsten particles, it can be discussed that there are two eddies in the melted domain: (1) in upwards at the upper

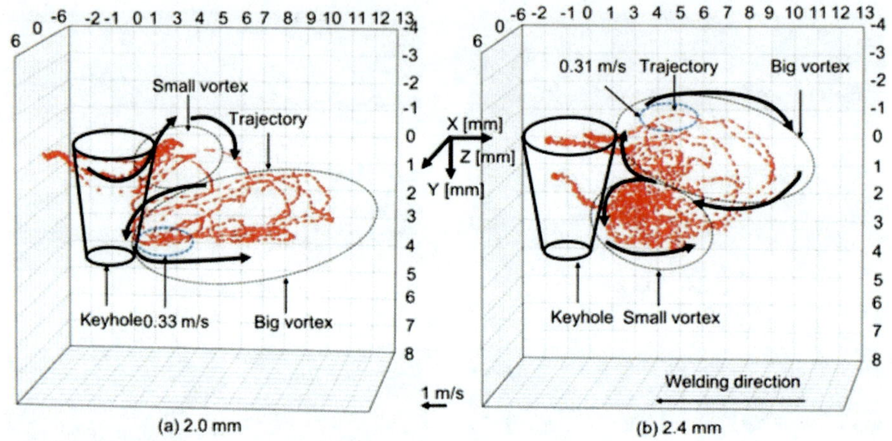

Fig. 3 Four-dimensional material flow at the melted domain

side of the weld pool and (2) in downwards at the lower side of the weld pool. As a result of this convection inside the melted domain, it has formed a vortex pair with upper vortex in accordance with the eddy at the upper side of the weld pool and lower vortex in accordance with the eddy at the lower side of the weld pool. In the case of 2.4 mm, when tracers move around the upper side, their velocity is strongly increased in upwards, and then in backward around rear kerf of the keyhole. After that, it moved down to a neutral place of workpiece before turning back to the rear kerf of the keyhole. When tracers located at the lower side of the weld pool, they were rapidly accelerated in downwards, and then in backward around the rear kerf of the keyhole. They were then transferred upwards to the middle part of the workpiece before turning back to the keyhole.

Based on the movement of tungsten particles, it can be noticed that two fluid vortices were formed in case of 2.4 mm. The vortex at the upper side of the weld pool is much stronger than the vortex at the lower side of the weld pool. In the case of 2.0 mm, when tracers located at the upper side, they moved in an upward tendency, and then in a backward tendency inside the weld pool and behind the keyhole. After that, it sunk to a neutral place of workpiece before coming back to the around keyhole. When tracers belonged at the lower side of the weld pool, they rapidly moved in downwards, and then in backward at the front kerf of the keyhole. They were then transported in upwards before turning back to the rear kerf of the keyhole. Consequently, two vortexes were formed and the lower vortex is much larger and stronger than the upper vortex. A large amount of plasma gas and plasma jet is pushed out in downward direction inside the keyhole with nozzle 2.0 mm. This results in a boarder vortex at the lower side of the weld pool. In the meantime, a major amount of plasma gas and plasma jet is blown along the melted domain surface and along the keyhole surface in upwards in case of 2.4 mm. This results in a larger vortex at the upper side of the weld pool. This convection is suitable with the change of the keyholes profile as expressed in Fig. 2. The highest speed of tracers at the melted domain is about 0.34 m/s at the lower side with nozzle 2.0 mm and is about 0.29 m/s at the upper side with nozzle 2.4 mm.

4 Conclusions

In this paper, the experiment on behavior of fluid vortex inside the melted domain was conducted. Some main results are drawn as follows:

Generally, two vortices are formed at the melted domain in PAW process. This mechanism is mainly based on the activation of the shear force due to the high plasma flow from the torch. The behavior of the shear force is principally decided based on the change of the keyhole contour. Therefore, the relative value of each vortex largely varies based on the change of welding parameters. This relative strength of each vortex is considered to decisively govern the heat transportation and material flow behavior at the motel domain, and also, the occurrence of the welding defects. The results in this investigation have enormous values to explode a comprehensive discussion about the relationship among the welding parameters with the material flow behavior and the crystallization process is not only for the PAW process but also for other welding and additive manufacturing processes.

References

1. McCaw RL (1979) Plasma arc welding of high-performance ship materials. David taylor naval ship research and development center report, SME-78/34
2. Vilkas EP (1991) Plasma arc welding of exhaust pipe system components. Weld J 70:49–52
3. Irving B (1997) Why aren't airplanes welded? Weld J 76:31–41
4. Zavadski A (2018) Advanced welding technologies used in aerospace industry. Doctoral thesis
5. Zhang YM, Zhang SB (1991) Observation of the keyhole during plasma arc welding. Weld J 2:53–58
6. Chen Y, Clark SJ, Huang Y et al (2021) In situ X-ray quantification of melt pool behaviour during directed energy deposition additive manufacturing of stainless steel. Mater Lett 286:129205
7. Leung CLA, Marussi S et al (2018) In situ X-ray imaging of defect and molten pool dynamics in laser additive manufacturing. Nat Commun 9:1355
8. Wu CS, Jia CB, Chen MA (2010) A control system for keyhole plasma arc welding of stainless steel plates with medium thickness. Weld J 11:225–231
9. Liu ZM, Wu CS, Liu YK, Luo Z (2015) Keyhole behaviors influence weld defects in plasma arc welding process. Weld J 9:281–290
10. Xiaoxia Jian C, Wu S (2015) Numerical analysis of the coupled arc-weld pool-keyhole behaviors in stationary plasma arc welding. Int J Heat Mass Transf 84:839–847
11. Van Anh N, Tashiro S, Van Hanh B, Tanaka M (2018) Behavior of exit keyhole diameter during switch off period in plasma keyhole arc welding. Adv Eng Forum 26:87–92

12. Van Anh N, Tashiro S, Van Hanh B, Tanaka M (2016) Visualization of weld pool convective flow in plasma keyhole arc welding. Front Appl Plasma Technol 9:1–6
13. Van Anh N, Tashiro S, Van Hanh B, Tanaka M (2017) Influence of pilot gas composition on convective pattern of weld pool surface in plasma keyhole arc welding. Q J Japan Weld Soc 35:98s–102s
14. Van Anh N, Tashiro S, Manh NH, Van Hanh B, Tanaka M (2020) Effect of the eddies formed inside a weld pool on welding defects during plasma keyhole arc welding. J Manuf Process 59:649–657
15. Van Anh N, Tashiro S, Van Hanh B, Tanaka M (2017) Development of plasma-MIG hybrid welding process. Q J Jpn Weld Soc 35:132s–136s
16. DebRoy T, Mukherjee T et al (2021) Metallurgy, mechanistic models and machine learning in metal printing. Nat Rev 6:48–68
17. Pham MS, Dogy B et al (2021) The role of side-branching in microstructure development in laser powder-bed fusion. Nat Commun 11:749
18. Kuril AA, Janaki Ram GD, Bakshi SR (2019) Microstructure and mechanical properties of keyhole plasma arc welded dual phase steel DP600. J Mater Process Technol 270:28–36
19. Geng W, Di Wu, Sun D, Li H, Che Y (2018) Microstructures and mechanical properties of plasma arc welded joints of ultra-high strength steel and aluminum alloy using Al–Si and Al–Cu fillers. ISIJ Int 58:1108–1116
20. Xu P, Jiang F, Meng S, Yi K et al (2018) Microstructure and mechanical properties of Al–Mg–Sc–Zr alloy variable polarity plasma arc welding joint. J Mater Eng Perform 27:4783–4790
21. Xu B, Tashiro S, Chen S, Jiang F, Van Anh N, Tanaka M (2020) Material flow analyses of high-efficiency joint process in VPPA keyhole flat welding by X-ray transmission system. J Clean Prod 250:119450
22. Xu B, Tashiro S, Jiang F, Chen C, Van Anh N, Tanaka M (2018) Numerical analysis of plasma arc physical characteristics under additional constraint of keyhole. Chin Phys B 27:034701
23. Manh NH, Van Anh N, Thanh Hai N et al (2020) Material flow behavior on weld pool surface in plasma arc welding process considering dominant driving forces. Appl Sci 10:3569
24. Van Anh N, Tashiro S, Manh NH, Le Anh H, Van Hanh B, Tannaka M (2020) Influence of shielding gas composition on molten metal flow behavior during plasma keyhole arc welding process. J Manuf Process 53:431–437
25. Van Anh N, Wu DS, Tashiro S, Tanaka M (2019) Undercut formation mechanism in keyhole plasma arc welding. Weld J 98:204–212

Experimental Study on Micro Milling of Glass

Ali Mamedov

1 Introduction

Glass is a widely used engineering material that is produced through the processing of silicon dioxide (SiO_2) compound, which is also frequently referred as silica. This material found a wide application in various industries, such as optics, biomedical, automotive and electronics, due to its sophisticated mechanical and physical properties. Various glass working and shaping processes, such as spinning, pressing, press-and-blow, blow-and-blow and casting, are used to produce common end products. All of these methods can be classified in three groups as discrete processes, continues processes and fiber making processes. However, due to the fact that in all above-mentioned processes glass is shaped in liquid or viscous state, and these methods have certain limitations. It would be beneficial to fabricate some glass parts through machining processes in solid state, but the brittle nature of glass makes it difficult. The machined surface is usually fractured and requires additional finishing processes that are costly and time consuming.

Recent developments in micro milling bring new flexibility in machining of various engineering materials including glasses. While investigating micro milling of brittle materials, such as glass, it is found that under certain cutting conditions brittle materials can be machined in a ductile regime. Machining in this regime results in a smooth surface with neglectable amount of minor cracks. On the other hand, machining under cutting conditions resulting in brittle cutting produces many cracks and results in poor surface quality. Despite numerous research in this field, it is still not exactly clear under what conditions transition between regimes occurs. Several researcher groups related transition of the cutting regime to different phenomena. Le Bourhis and Rouxel [1] stated that transition is related to fracture

A. Mamedov (✉)
College of Engineering and Technology, American University of the Middle East,
Egaila, Kuwait
e-mail: ali.mamedov@aum.edu.kw

R. K. Agarwal (ed.), *Recent Advances in Manufacturing Engineering and Processes*,
Lecture Notes in Mechanical Engineering,
https://doi.org/10.1007/978-981-16-3934-0_11

toughness of glass along with pressure and temperature formed in cutting shear zone. Moriwaki et al. [2] related transition to the undeformed chip thickness and stated that if the chip thickness is below a critical value machining is ductile. However, when the chip thickness reaches a critical value brittle fracture occurs and its effect increases with increasing chip load. Rouxel and Sangleboeuf [3] showed that the glass can experience material softening under loading at elevated temperatures, which affects mechanical properties, such as fracture toughness, hardness and elastic modulus. Komanduri [4] showed that cutting mechanism can be shifted from brittle fracture to ductile by applying hydrostatic stress and/or temperature. Consequently, thermal softening of the brittle material with laser was studied by Hwang et al. [5] and effect of hydrostatic pressure and temperature on the cutting regime transition via ultrasonic vibration was studied by Zhou et al. [6]. The practical application of the micro ball-end milling process of the glass was also demonstrated by Matsumura and Ono [7]. Where authors showed that crack-free machining of the glass is possible with axial depth of cut of much larger than a micrometer. Liu et al. [8] evaluated the effect of the inclination angle of the spindle on the surface quality and stated that 45° inclination angle resulted in the highest quality of the machined surface. Arif et al. [9, 10] analyzed the critical conditions for different modes of machining and established a relationship between the radial depth of cut and the depth of subsurface damage generated during brittle machining regime. Authors also evaluated a critical feed per tooth value that will result in removal of machining induced cracks from final machined surface by the consecutive cut.

This study offers an experimental evaluation of micro milling of glass material under various cutting conditions and understanding of the effect of cutting speed, federate and axial depth of cut on generated surface quality.

2 Methodology and Experiments

The quality of the generated surface is highly affected by the cracks formed during machining of the glass. Therefore, the evaluation of the crack formation mechanisms is essential for understanding of the surface generation. Literature survey shows that crack formation during brittle regime mostly occurs in the region of the high-pressure zone formed by the tool tip in the chip formation zone. Thus, these cracks can propagate to the final generated surface or can be removed by consecutive cut. Patten et al. [11] mentioned that this kind of cracks can be minimized by controlling cutting conditions resulting in reduced stress formation in front of cutting edge.

Another possible crack formation may occur on the machined bottom surface. Cutting edge applies a local load on the bottom surface and once the tool passes the loading on the surface is released, however, the residual stress at this boundary can initiate lateral cracks. Similarly, these cracks can propagate to the final generated surface or can be removed by consecutive cut. Both crack formation types are schematically shown in Fig. 1.

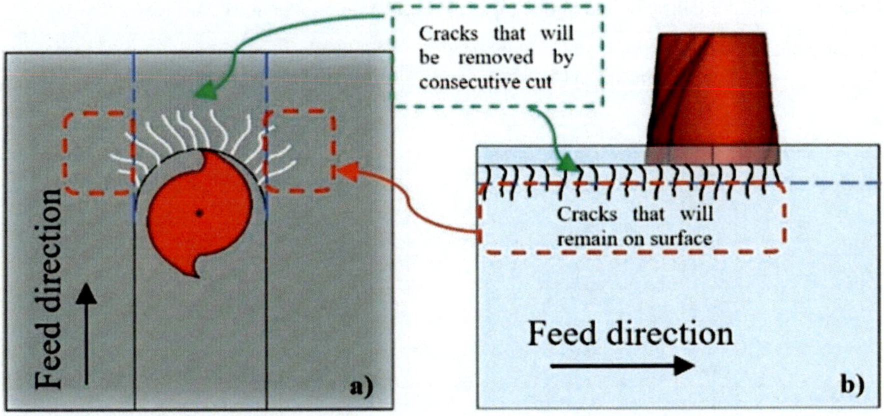

Fig. 1 Schematic representation of **a** Cracks formed in front of the cutting edge; **b** Lateral cracks formed on the bottom surface

Experiments were performed on glass microscope slides using 500-μm diameter two fluted tungsten carbide micro end mill, and surface quality was evaluated using optical microscope.

In order to evaluate effect of the cutting conditions micro milling experiments at different cutting speed, federate and depth of cut were performed. Cutting conditions for all experiments and generated surfaces evaluated under optical microscope are presented in Table 1.

The quality of the surface was analyzed using optical microscope and evaluated using linguistic terms, where "very high" corresponds to surface with almost no cracks and "very low" corresponds to surface with extremely high amount of cracks.

In order to present more meaningful result evaluation, feed per tooth and depth of cut parameters can be evaluated together in a form of an amount of the material removed by single cutting flute in one rotation [12].

From the results, it can be clearly seen that generated surface quality drastically decrease with increase in the amount of removed material. Lower depth of cut and feedrate result in less amount of cracks formed on machined surface. A comparison of two extreme cases under same cutting speed is presented in Fig. 2.

Also, it observed that the increase cutting speed results in better surface quality. Liu et al. [8] related it to increase in the temperature in cutting area that result in thermal softening of the glass. However, extremely high-cutting speed will also have a negative effect on surface quality. Therefore, an optimum value of cutting speed for brittle materials should be determined. A comparison of two cases under with same feed per tooth and depth of cut is presented in Fig. 3.

From micro milling experiments of glass, it was observed that selection of right cutting parameters is extremely important and it has direct effect on cutting mechanism. If cutting parameters are selected right glass starts to act as a ductile

Table 1 Cutting conditions and surface quality evaluation of machined surfaces

Experiment #	Spindle speed (rpm)	Cutting speed (m/min)	Feed/tooth (μm/tooth)	Depth of cut (μm)	Surface quality
1	3200	5	5	50	Very low
2	3200	5	5	25	Low
3	3200	5	2.5	25	Low
4	3200	5	2.5	10	Medium
5	3200	5	1	10	Medium
6	5000	7.85	5	50	Very low
7	5000	7.85	5	25	Very low
8	5000	7.85	2.5	25	Low
9	5000	7.85	2.5	10	High
10	5000	7.85	1	10	Very high
11	6400	10	5	50	Very low
12	6400	10	5	25	Very low
13	6400	10	2.5	25	High
14	6400	10	2.5	10	High
15	6400	10	1	10	Very high

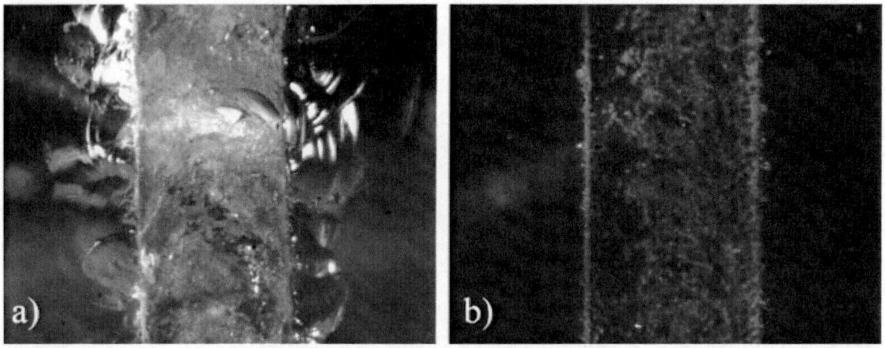

Fig. 2 Generated surface for **a** Experiment 6 and **b** Experiment 10

material and crack formation due to chipping and brittle breakage of the glass is minimized. In order to further validate presented experimental study, a micro channel similar to one used in lab-on-a-chip applications was manufactured using following cutting parameters, which resulted in high-surface quality, as shown in Fig. 4.

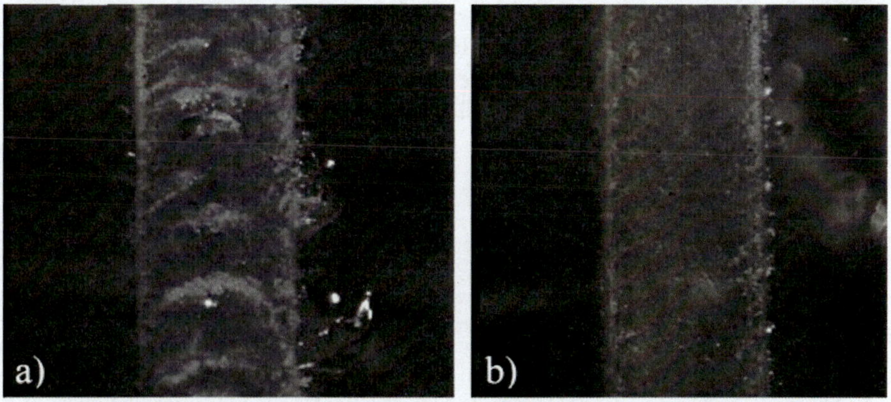

Fig. 3 Generated surface for **a** Experiment 3 and **b** Experiment 13

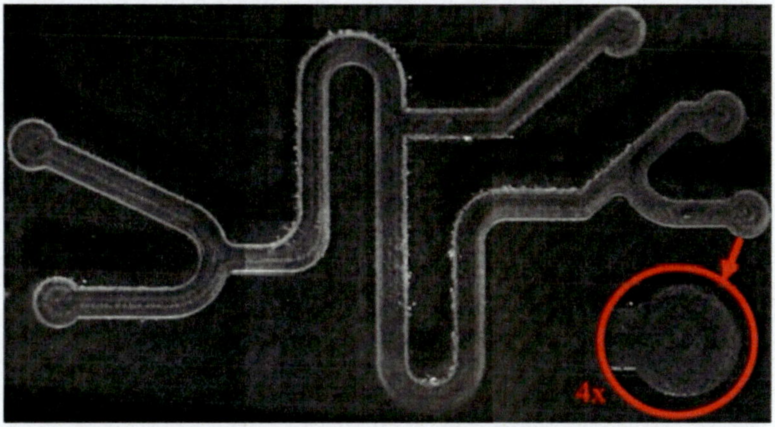

Fig. 4 Machined micro channel [cutting speed 10 m/min, feed per tooth 1 μm/tooth and depth of cut 10 μm]

3 Conclusion

Glass material is considered as difficult process because of its tendency toward brittle fracture during processing. Therefore, a substantial research was performed in order to achieve a ductile mode machining and avoid brittle fracture. The plastic deformation is dominant mechanism of material removal in ductile mode machining, therefore, any cracks produced due to possible fracture in the cutting zone are prevented from extending into the machined surface. Experiments showed that this can be achieved by selection of right cutting parameters. Experiments showed that the generated surface quality drastically decrease with increase in the amount of removed material. Lower depth of cut and feedrate result in less amount of cracks formed on machined surface. Also, it observed that the increase cutting speed results in better surface quality. However, extremely high-cutting speed will also have a negative effect on surface quality. Therefore, an optimum value of cutting speed for brittle materials should be determined.

References

1. Le Bourhis E, Rouxel T (2003) Indentation response of glass with temperature. J Non-Cryst Solids 316:153–159
2. Moriwaki T, Shamato E, Inoue K (1992) Ultraprecision ductile cutting of glass by applying ultrasonic vibration. Annals of CIRP 41:141–144
3. Rouxel T, Sangleboeuf J (2000) The brittle to ductile transition in a soda-limesilica glass. J Non-Cryst Solids 271:224–235
4. Komanduri R (1996) On material removal mechanisms in finishing of advanced ceramics and glasses. Annals of CIRP 45:509–513
5. Hwang D, Choi T, Grigoropoulos C (2004) Liquid-assisted femtosecond laser drilling of straight and three-dimensional microchannels in glass. Appl Phys A 79:605–612
6. Zhou M, Ngoi B, Yusoff M, Wang X (2006) Tool wear and surface finish in diamond cutting of optical glass. J Mater Process Technol 174:29–33
7. Matsumura T, Ono T (2005) Glass machining with ball end mill. Trans NAMRI/SME 33:319–326
8. Liu HT, Sun YZ, Shan DB (2013) Experimental research of brittle-ductile transition conditions and tool wear for micromilling of glass material. Int J Adv Manuf Technol 68:1901–1909
9. Arif M, Rahman M, San WY (2012) Analytical model to determine the critical conditions for the modes of material removal in the milling process of brittle material. J Mater Process Technol 212:1925–1933
10. Arif M, Rahman M, San WY (2010) Analytical model to determine the critical feed per edge for ductile-brittle transition in milling process of brittle materials. Int J Mach Tools Manuf 51:170–181
11. Patten J, Cherukuri H, Yan J (2019) Ductile-regime machining of semiconductors and ceramics. High-Pressure Surf Sci Eng:543–632 (chapter 6)
12. Mamedov A, Lazoglu I (2016) An evaluation of micro milling chip thickness models for the process mechanics. Int J Adv Manuf 87:1843–1849

The Effect of Electrode Tip Diameter on Indentation Feature and Nugget Diameter of Resistance Spot Welded Automotive Steel Joint

Hee-Seon Bang, Kyoung-Hak Kim, Jong-Hee Kim, Kyung-Hwan Oh, and Jin-Tae Jeong

1 Introduction

Resistance spot welding (RSW) is the dominant welding technology used in automotive industry for production of the car body shells as the process is quite flexible with higher productivity and can be robotised easily [1–4]. Weld surface appearance and size of the nugget are the essential requirements in RSW. Different process conditions (e.g. welding current, weld time and electrode force) simultaneously control the formation of weld nugget, and indentation marks on the workpiece surfaces [5–7]. In general, highelectrode pressure is applied on sheet surfaces during RSW to contain the molten nugget. As a result, electrodes leave indentation marks on the weld surface around 10–20% of the sheet thickness. These indentation marks adversely affect the appearance of the weld surface and additionally act as a stress riser that markedly reduce the joint shear strength during loading. The lower indentation depths could potentially cut down the cost associated with the subsequent processing of the welded parts prior to coating. The indentation depth can be reduced by increasing the contact area of the electrode surface such as using a smooth and flat electrode or by employing a copper backing plate at the face side of the joints [8]. However, increase in flat electrode diameters tends to decrease the weld nugget size, which in turn reduces the joint strength [9]. Alternatively, literature on RSW suggested applying higher welding currents to achieve an appropriate nugget diameter with adequate strength [10]. In the present work, cold rolled steel sheets (SPCC) are joined by RSW process by employing a

H.-S. Bang (✉) · K.-H. Kim
Department of Welding and Joining Science Engineering, Chosun University, Gwangju 61452, Republic of Korea
e-mail: banghs@chosun.ac.kr

J.-H. Kim · K.-H. Oh · J.-T. Jeong
Department of Welding and Joining Science Engineering, Graduate School, Chosun University, Gwangju 61452, Republic of Korea

R. K. Agarwal (ed.), *Recent Advances in Manufacturing Engineering and Processes*, Lecture Notes in Mechanical Engineering,
https://doi.org/10.1007/978-981-16-3934-0_12

flat copper backing plate at the lower electrode side to achieve an elegant joint appearance along with lower indentation depth. Further, the influence of upper electrode diameter and welding currents on indentation depth are investigated through mechanical and macro characterisation.

2 Experimental Procedure

The SPCC of a 1.2 mm and a 1.6 mm thick were selected as a base metal, respectively. This steel is widely used for roof, fender, hood and oil pans in automobiles. Table 1 lists the chemical compositions and mechanical properties of the base metal. The specimen dimensions of 1.2 and 1.6 mm thick steel sheets were 160 mm × 50 mm. Single phase pneumatically operated resistance spot welding machine was employed to join specimens in lap configuration. Chromium-copper (Cr–Cu) made dome cap type electrode with cap diameter of 16 mm and flat contact diameter of 6 mm was used as a lower electrode. In case of face side electrodes, dome cap of 16 mm diameter with different flat tip diameters such as 6 and 8 mm were used for experiments. A flat copper backing plate of 10 mm in thickness, 110 mm in length and 30 mm in width was positioned between the lower electrode and steel sheet. Figure 1a shows the schematic diagram of the experimental setup. The indentation depth was measured at the centre of the electrode indentation using optical microscope (OM) and dital gauge as a depth of the imprint from top surface of the sheet. Table 2 shows the welding conditions used in RSW of SPCC steel sheets. The cross section of weld nuggets was measured after polishing and etching with 3% Nital solution. A Shimatzu EHF-EG200KN-40L universal tensile testing machine was used to evaluate the tensile shear load of the welded joint. Micro Vickers hardness test was conducted as per KS B ISO 14273 standard. Figure 1b shows the Vickers hardness measured locations across the joint cross section (In total, 7 locations at the nugget, 4 locations at the heat affected zone and 6 locations at the base metal).

Table 1 Chemical compositions and mechanical properties of base metal

Chemical compositions (wt%)					
Material	C	Si	Mn	P	S
SPCC	0.12	0.002	0.5	0.05	0.051

Mechanical properties			
Material	Yield strength (MPa)	Tensile strength (MPa)	Elongation (%)
SPCC	162	292	50

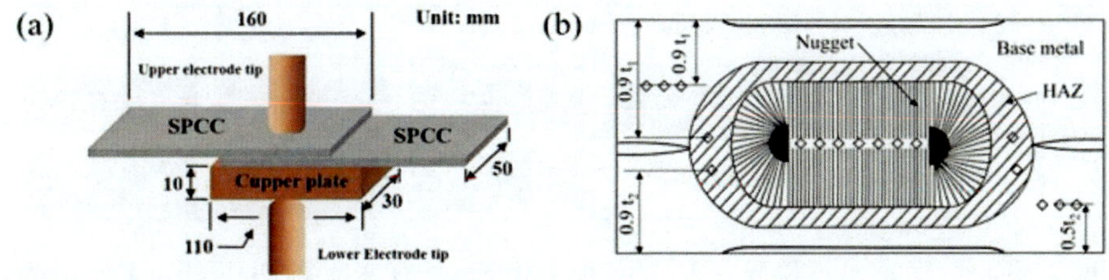

Fig. 1 Schematic diagram of experimental setup for RSW

Table 2 Welding conditions for RSW of SPCC sheets

Conditions	1.2–1.2 mm SPCC sheets	1.6–1.6 mm SPCC sheets
Pressing time (ms)	600	600
Current 1 (kA)	9, 9.5, 10, 10.5 and 11	7
Weld time 1 (ms)	600	250
Current 2 (kA)	–	9, 9.5, 10, 10.5 and 11
Weld time 2 (ms)	–	600
Cooling time (ms)	200	
Upper electrode tip diameter (mm)	6, 8	
Lower electrode tip diameter (mm)	6 (fixed)	

3 Results and Discussion

3.1 Nugget Characteristics

Figure 2 indicates the macroscopic view of nuggets in spot welding of 1.2–1.2 mm thick SPCC steel sheets at different welding currents from 9 to 11 kA for upper electrode tip diameter (ETD) of 6 (ϕ6) mm and 8 (ϕ8) mm, respectively. Nuggets at interface were symmetrically formed under all welding currents using an ETD of ϕ6 mm. This was attributed to symmetrical heat generation during heating and cooling cycles. An ETD of ϕ8 mm did not develop the nuggets at welding currents ranging from 9 to 9.5 kA due to insufficient heat generation. Further increase in welding current above 9.5 kA increased the heat input, which facilitated the nugget formation. However, for the upper ETD of ϕ10 and ϕ16 mm exhibited insufficient nugget size due to incomplete fusion. This was attributed to the significant reduction in heat generation due to decrease in current density at larger electrode diameter. As shown in Fig. 3, for ETD of ϕ6 mm, nuggets were generated at all welding currents between 9 and 11 kA, when the SPCC thickness increased from 1.2 to 1.6 mm. In case of the ϕ8 mm upper ETD, the nugget only developed at a welding current of 11 kA. At lower currents ranging from 9.0 to 10.5 kA, the

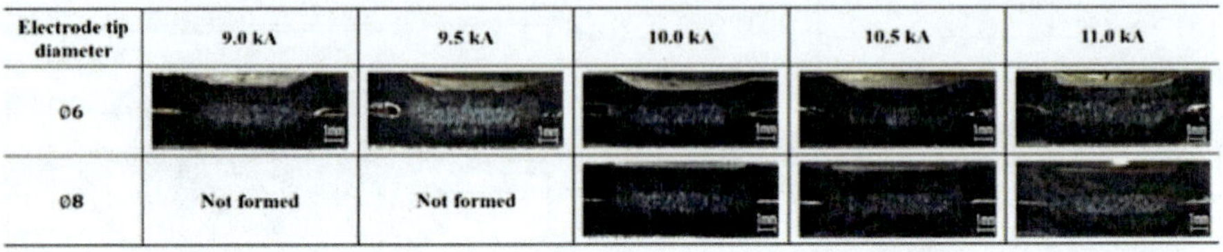

Fig. 2 Shape of nugget in RSW of 1.2 mm thick SPCC sheets using ETDs of ϕ6 and ϕ8 mm at different welding currents

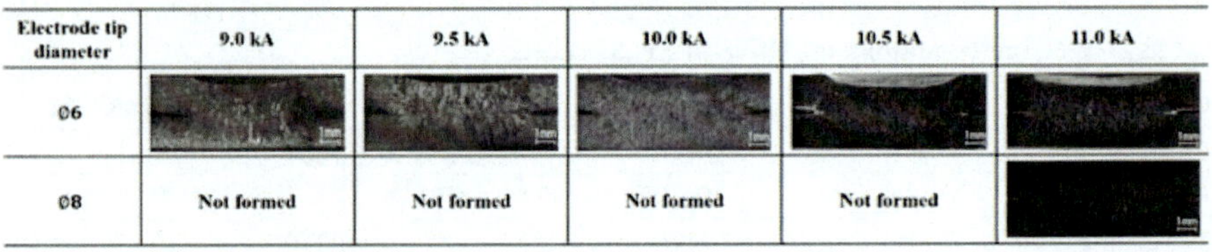

Fig. 3 Shape of nugget in RSW of 1.6 mm thick SPCC sheets using ETDs of ϕ6 and ϕ8 mm at different welding currents

nugget was not formed by the reduction of heat generation at the interface in thicker sections. Higher ETDs of ϕ10 and ϕ16 mm resulted in inappropriate nugget shape that could be attributed to significantly reduced heat generation due to increase in ETD.

3.2 *Indentation Depth*

Figure 4 shows the measured nugget size and indentation depth for upper ETDs of ϕ6 and ϕ8 mm at different welding currents in spot welding of 1.2 and 1.6 mm thick SPCC steel sheets, respectively. In RSW of 1.2 mm thick steel sheets using ETD of ϕ6 mm, Fig. 4a exhibits the increased weld nugget size ranging from 5.5 to 6.19 mm, when the welding current increased from 9 to 11 kA. Similarly, with ETD of ϕ8 mm, the nugget size increased from 6.32 to 6.68 mm, when the welding current increased from 10 to 11 kA. In RSW of 1.6 mm thick steel sheets using ETD of ϕ6 mm, an increase in welding current from 9 to 11 kA led to an increase in nugget size ranging from 7.11 to 7.32 mm. And the nugget size formed at welding current of 11 kA with ETD of ϕ8 mm was 7.1 mm. Figure 4a illustrates that nugget size increases with an increase in welding current at any given electrode diameter, which is attributed to an increase in heat generation at higher welding current. Further, all weld nuggets achieved the minimum nugget size limit of $5\sqrt{\text{tickness}}$, i.e. 5.48 and 6.32 mm for 1.2 and 1.6 mm thick steel sheets, respectively. Figure 4b depicts an increase in upper indentation depth from 0.44 to 0.51 mm with an increase in welding current from 9 to 11 kA using an upper ETD

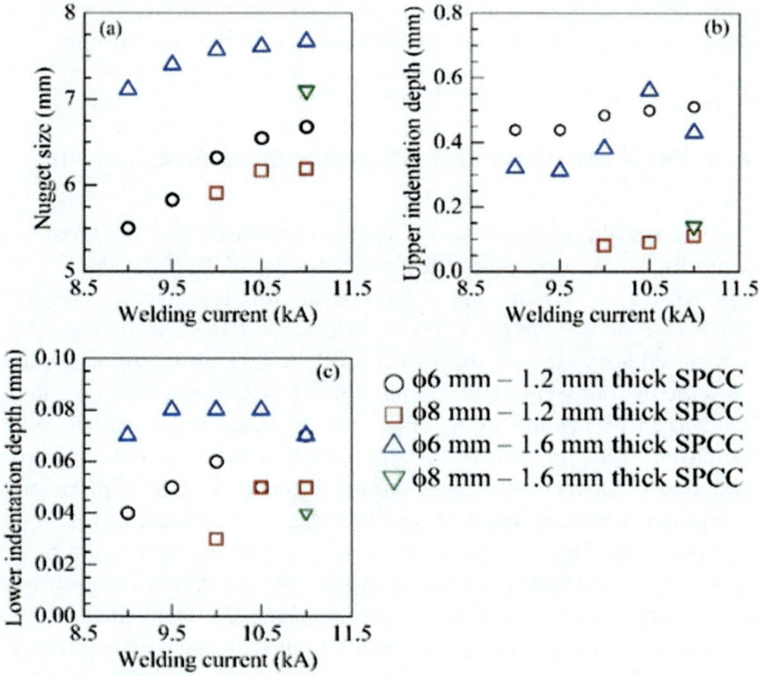

Fig. 4 Measured **a** nugget sizes, **b** upper indentation depths and **c** lower indentation depths in RSW of SPCC sheets using upper ETDs of ϕ6 and ϕ8 mm

of ϕ6 mm in spot welding of 1.2 mm SPCC steel sheets. In RSW of 1.2 mm SPCC sheets using ETD of ϕ8 mm, the upper indentation depths increased from 0.09 to 0.11 mm, when welding current increased from 10 to 11 kA. A 1.6 mm thick SPCC welded joint using ETD of ϕ6 mm exhibited the upper indentation depth ranging from 0.32 to 0.58 mm. In case of ETD of ϕ8 mm, upper indentation depth was 0.14 mm. The sheet separation phenomenon that degrades the weld strength and disrupts the surface morphology for both thicknesses of steel sheets was observed in welded joint produced with ETD of ϕ6 mm, as shown in Figs. 2 and 3. This was attributed to a smaller ETP and excessive welding force. As a result, the indentation depth was significantly reduced by using ETD of ϕ8 mm. Figure 4c shows the lower side indentation depth, which was measured from 0.04 to 0.07 mm with variation of welding current from 9 to 11 kA in spot welding of 1.2 mm thick SPCC sheets using upper ETD of ϕ6 mm. Similar to ETD of ϕ6 mm, the lower indentation depth, which had ranges of 0.03–0.05 mm, also increased with increasing welding current from 10 to 11 kA in spot welding of 1.2 mm SPCC sheets using ETD of ϕ8 mm. The lower indentation depth was formed from 0.07 to 0.08 mm, when an ETD of ϕ6 mm was used to weld the 1.6 mm thick SPCC sheets. In case of ETD of ϕ8 mm, the lower indentation depth was 0.04 mm. A comparison of Fig. 4b, c shows that lower values of indentation depth at the weld bottom surface in comparison to that of the upper side of the weld, which is due to the presence of flat copper plate at the bottom side of the welded joint.

3.3 Tensile Shear Load and Microhardness Distribution

Figure 5 indicates the comparison of tensile shear load and fractured specimen morphology for the welded joint of 1.2 and 1.6 mm thick SPCC sheets at welding currents of 10 and 11 kA using upper ETD of ϕ8 mm, respectively. The maximum tensile shear load of 1.2 mm thick SPCC welded joint was approximately 6.55 kN and the fractured specimen exhibited tear mode failure in optimised welding conditions. In case of spot welding of 1.6 mm thick SPCC sheet, the maximum tensile shear load was 11.28 kN and the specimen was fractured as plug mode failure. The result of tensile shear test shows that the tensile shear load obtained from both welded joints was satisfied as Korean Industrial Standard (KS) requirements. Note that the required minimum value of tensile shear load presented in KS is 6.1 and 9.2 kN, respectively. Higher weld strength in spot welding of 1.6 mm thick SPCC sheets was due to the larger nugget size, as shown in Fig. 4a. Figure 6 represents the micro Vickers hardness profiles measured along the weld traverse section in optimised indentation depth condition for the joint of 1.2 and 1.6 mm thick SPCC sheets, respectively. A similar trend was observed in the hardness profile for two different thicknesses of steel sheets. As shown in Fig. 6, the hardness values are observed to increase from base metal to the weld nugget that can be attributed to the change in microstructure of the welded joint. The minimum hardness was measured in the base metal for 1.2 and 1.6 mm thick steel sheets as the microstructure of base metal consists of ferrite. The nugget zone exhibited the maximum hardness of 135–145 Hv and 112–145 Hv for spot welding of 1.2 and 1.6 mm SPCC steel sheets, respectively. The values of nugget zone hardness were nearly double to that of the base metal hardness that could be attributed to the formation of lathe martensite in

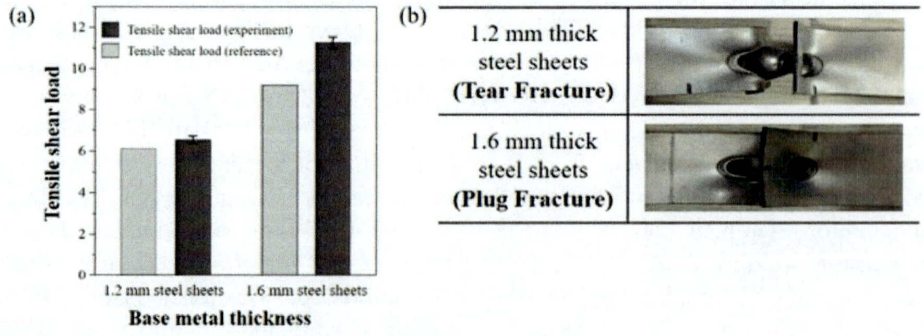

Fig. 5 Comparison of tensile shear strength and fracture specimen for the weld of 1.2 and 1.6 mm thick SPCC sheets at upper ETD of ϕ8 and welding current of 10 and 11 kA

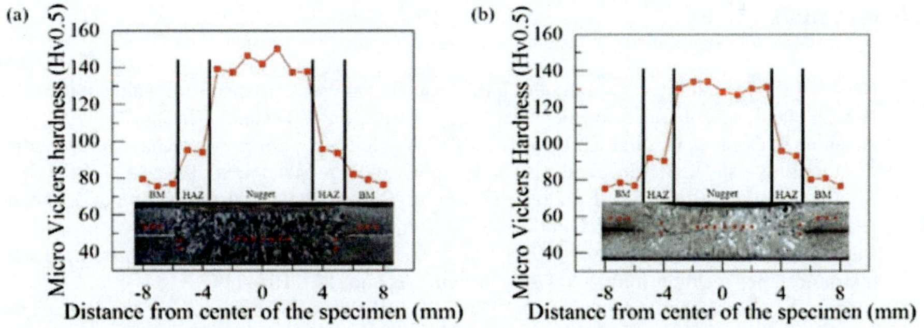

Fig. 6 Distribution of microhardness along weld transverse section for the spot welding of **a** 1.2 mm thick SPCC sheets at welding current of 10 kA and ETD of ϕ8 mm, and **b** 1.6 mm thick SPCC sheets at welding current of 11 kA and ETD of ϕ8 mm

nugget zone owing to higher cooling rates in RSW. Similar trend of hardness profile was reported earlier by several researchers in RSW of SPCC sheets [11, 12]. Therefore, it is concluded that this technique suggested in present study is effective in reducing the indentation depth in RSW.

4 Conclusions

In present study, resistance spot welding of automotive SPCC sheets has been successfully performed and the effect of upper ETD on formation of indentation depth and nugget size has been examined. The following conclusions are derived:

(1) The present study illustrates the effect of ETD on indentation characteristics in resistance spot welding of 1.2–1.2 mm and 1.6–1.6 mm thick SPCC sheets. The limit of minimum nugget size is obtained when spot welding is carried out by upper flat ETDs of 6ϕ and 8ϕ mm.
(2) The minimum indentation depth is achieved using an upper ETD of ϕ8 mm in spot welding of 1.2–1.2 mm and 1.6–1.6 mm thick SPCC sheets at welding currents of 10.0 and 11.0 kA, respectively.
(3) The maximum tensile shear load of 1.2 mm thick SPCC sheets was 6.55 kN under welding current of 10 kA. The maximum tensile shear load of 1.6 mm thick SPCC sheets reached 11.28 kN under welding current of 11 kA.

Acknowledgements This paper is supported by Korea Institute for Advancement of Technology (KIAT) grant funded by the Korea Government (MOTIE) (P0008425, The Competency Development Program for Industry Specialist).

References

1. Chao YJ (2003) Ultimate strength and failure mechanism of resistance spot weld subjected to tensile, shear, or combined tensile/shear loads. J Eng Mater Technol 125(2):125–132
2. Aslanlar S, Ogur A, Ozsarac U, Ilhan E (2008) Welding time effect on mechanical properties of automotive sheets in electrical resistance spot welding. Mater Des 29(7):1427–1431
3. Ma C, Chen DL, Bhole SD, Boudreau G, Lee A, Biro E (2008) Microstructure and fracture characteristics of spot-welded DP600 steel. Mater Sci Eng A 485(1–2):334–346
4. Luo X, Ren J, Li D, Qin Y, Xu P (2016) Macro characteristics of dissimilar high strength steel resistance spot welding joint. Int J Adv Manuf Technol 87:1105–1113
5. Mazur W, Kyriakopoulos A, Bott N, West D (2016) Use of modified electrode caps for surface quality welds in resistance spot welding. J Manuf Process 22:60–73
6. Williams NT, Parker JD (2004) Review of resistance spot welding of steel sheets part 1 modelling and control of weld nugget formation. Int Mater Rev 49(2):45–75
7. Aslanlar S (2006) The effect of nucleus size on mechanical properties in electrical resistance spot welding of sheets used in automotive industry. Mater Des 27(2):125–131
8. Berezienko VP, Furmanov SM (2004) Reducing the depth of indentation from electrodes in resistance spot welding. Weld Int 18(2):39–144
9. O'Brien, RL (1991) Welding handbook. Am Weld Soc 2(8):532–552
10. Zhao YY, Zhang YS, Lai XM, Wang PC (2013) Resistance spot welding of ultra-thin automotive steel. J Manuf Sci Eng 132(2):021012–021021
11. Pouranvari M, Marashi SPH (2010) On the failure of low carbon steel resistance spot welds in quasi-static tensile-shear loading. Mater Des 31(8):3647–3652
12. Nghiem NQ, Hwang HY, Chen JS (2012) Correlation of hardness with mechanical properties of spcc steel spot weld. Appl Mech Mater 157–158:1404–1409

Product Design and Intelligent Manufacturing

Predictive Maintenance Using Recurrent Neural Network Without Feature Engineering

F. Chalvin, Y. Miyamae, Y. Oku, and K. Nakahara

1 Introduction

Machine health monitoring as an enabler of predictive maintenance for so-called industry 4.0 is an active area of research that seems to give good results when paired with machine learning techniques. Being able to carry out maintenance at just the right time, avoiding machine downtime due to an unplanned breakdown while maximizing the lifetime of the components, is of great interest for industrial actors as it leads to substantial savings from increased productivity and reduction of replacement parts costs. Classically, various statistical methods were employed in order to either predict the remaining time before the next necessary maintenance operation [1] or simply to carry out preventive maintenance at regular intervals based on the usual failure rate of the different machine parts [2]. Numerous studies were carried out to optimize the schedule of the maintenance in regard to the duration of the production steps [3, 4]. With the advances in machine learning, techniques such as support vector machines (SVMs) [5], k-nearest neighbors (k-NN) [6] and more recently neural networks [7] were tried with various amount of success. Due to the time-based nature of the data collected from live machinery, recurrent neural networks (RNN) are good candidates for machine health prediction as shown by studies using ConvLSTM [8] or a simple stacked RNN [9]. Another challenge in machine health monitoring arises from the fact that machines are more often working fine than broken. This situation leads to a large imbalance between the anomalous and non-anomalous (healthy) classes in the data collected. To counter this problem, various techniques have been developed such as oversampling or undersampling [10], boosting [11], and cost-sensitive learning [12].

In this paper, we will present the results of an encoder–decoder RNN architecture using long short-term memory cells (LSTM) [13] to carry out machine health pre-

F. Chalvin (✉) · Y. Miyamae · Y. Oku · K. Nakahara
Rohm Co. LTD., Interdisciplinary Research Centre, Kyoto 615-8585, Japan
e-mail: florian.chalvin@dsn.rohm.co.jp

R. K. Agarwal (ed.), *Recent Advances in Manufacturing Engineering and Processes*, Lecture Notes in Mechanical Engineering,
https://doi.org/10.1007/978-981-16-3934-0_13

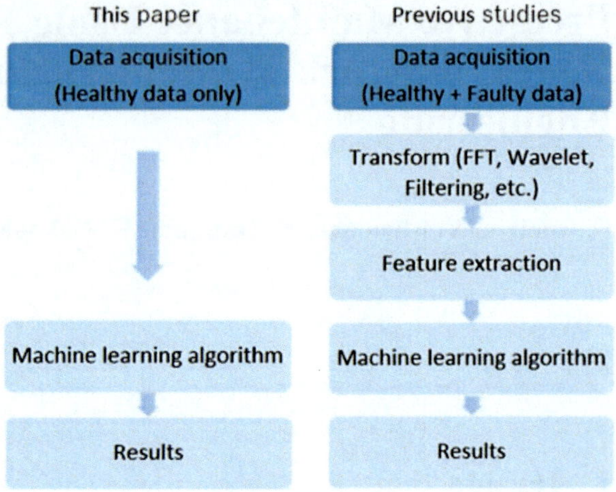

Fig. 1 Comparison between the usual method (right) and the method suggested in this paper (left)

dictions and detect upcoming mechanical failures. The encoder–decoder architecture has not yet been used to carry out such predictions. Contrary to previous techniques which rely on complex data preprocessing or statistical analysis for feature engineering, our method uses unsupervised learning to train the RNN from raw healthy data only. This solves the data imbalance problem by removing the need to train on anomalous data points and allows for a simpler data processing pipeline. These improvements are especially interesting for machine health monitoring on critical systems where one cannot afford to let the machine run to failure in order to gather anomalous data. A simpler data processing pipeline is also interesting for edge applications where computing power is limited. A diagram shown in Fig. 1 summarizes the difference between the proposed and conventional methods.

We tested this architecture on two different datasets, one made of wafer cutting machines' blades data collected in our company and the other being the NASA IMS dataset [14] resulting from bearings reliability tests. The results obtained on both these datasets show that one can adapt the method to various situations with limited overhead. The rest of the paper is organized as follows: Sect. 2 introduces the datasets and explains the procedure used to split the data into training and test sets. Section 3 gives a brief overview of the LSTM cells theory before presenting the encoder–decoder architecture and how it was implemented. Section 4 details the results obtained in the experiments and a brief conclusion regarding the limits of this technique is given in Sect. 5.

2 Data Environment

This section details the type of data used for the experiments. First, our internal dataset with data collected on wafer cutting machines. Predicting the remaining lifetime of the blade of wafer cutting machines is especially important as a failure during operation carries a risk of breaking the wafer and wasting the chips produced in addition to the cost of the machine downtime. The second dataset used is the open IMS bearings dataset [14] which has already been used for various studies regarding bearings lifetime predictions [15, 16]. Bearings are ubiquitous in machinery; therefore, being able to predict bearing failures reliably is very interesting for smooth operation of factories.

2.1 Wafer Cutting Dataset

Wafer cutting machines are equipped with two electric motors that are used to drive the blades. We have access to three sensors: one current sensor for each motor and the last one recording the vibrations of the table holding the wafer. A simplified schematic of the wafer cutting machine is shown in Fig. 2. Data are continuously sampled at 20 kHz. As mentioned earlier, collecting anomalous data is challenging as the machines are usually running in their normal state. In our situation, there are only two observed instances of blade breaking in the several months' worth of data collected. As a consequence, it is necessary that the RNN is able to detect the anomalous states while training on healthy data only. To train the RNN, we used data randomly sampled in a timeframe where we knew the system was not likely to be damaged as summarized in Fig. 3. Figure 4 shows an example of data collected from the graph we can observe that the stage vibration data have a wide range of values and are quite noisy, whereas motors current data are very clean. In total, we used 6400 training examples representing about 26 min of machine operation. The structure of these examples is explained in Sect. 3.2 and a complete example is shown in Fig. 8.

2.2 IMS Bearings Dataset

From the dataset description [14]:

> "Four bearings were installed on a shaft. The rotation speed was kept constant at 2000 RPM by an AC motor coupled to the shaft via rub belts. A radial load of 6000 lbs is applied onto the shaft and bearing by a spring mechanism. PCB 353B33 High Sensitivity Quartz ICP accelerometers were installed on the bearing housing, one accelerometer for each bearing. Each data set consists of individual files that are 1-second vibration signal snapshots recorded at specific intervals. Each file consists of 20,480 points with the sampling rate set at 20 kHz."

For our experiments, we used the 2nd dataset contained in this data packet. Figure 5 shows the data provided by the four accelerometers as well as the split between the training and test data. In this dataset, the experiment ended with an outer race failure of bearing number 1. Once again anomalous data are rare and would not be available beforehand on an actual system anyway; therefore, we focus

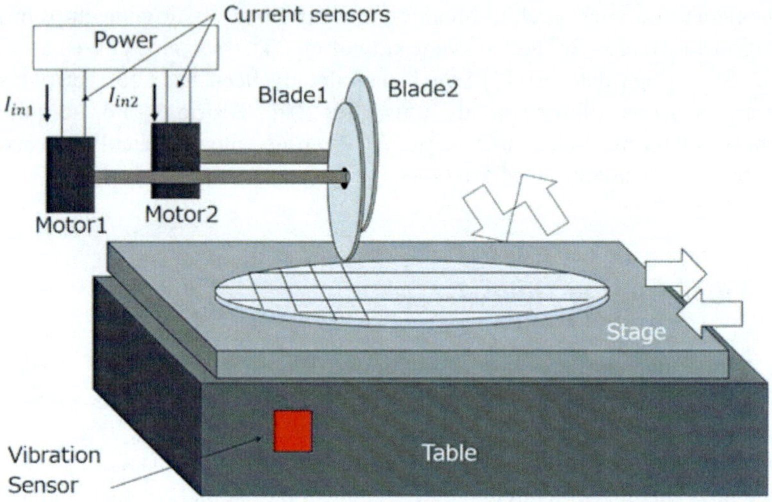

Fig. 2 Wafer cutting machine schematic and sensors locations

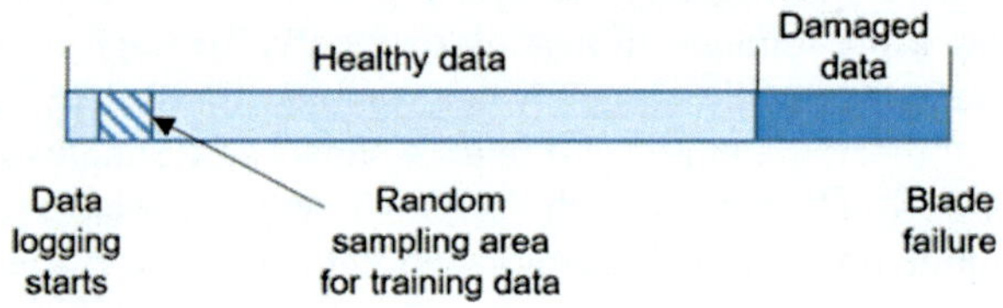

Fig. 3 Data repartition for the wafer cutting dataset. Training data were sampled near the beginning of the recording where the blade is assumed not to be damaged. Data after the last training sample were used as test set

on training only from healthy data obtained at the beginning of the experiment. We took the 171 first files of the dataset as training examples leaving the 813 others as test data. In real-time, this corresponds to roughly the 28 first hours of the 164 h that the experiment lasted for, giving us around 6800 training examples in total. The structure of these examples is explained in Sect. 3.2, and a complete training example is shown in Fig. 8. This split was arbitrarily chosen in order to have a sufficiently large training set while staying reasonably far from the anomalous region in order to avoid contamination in the training data.

3 Neural Network Architecture

To carry out our experiments, we used an encoder–decoder architecture [17]. This architecture has obtained good results in automated translation tasks [18] and speech-related applications [19] when using RNN. When using convolutional

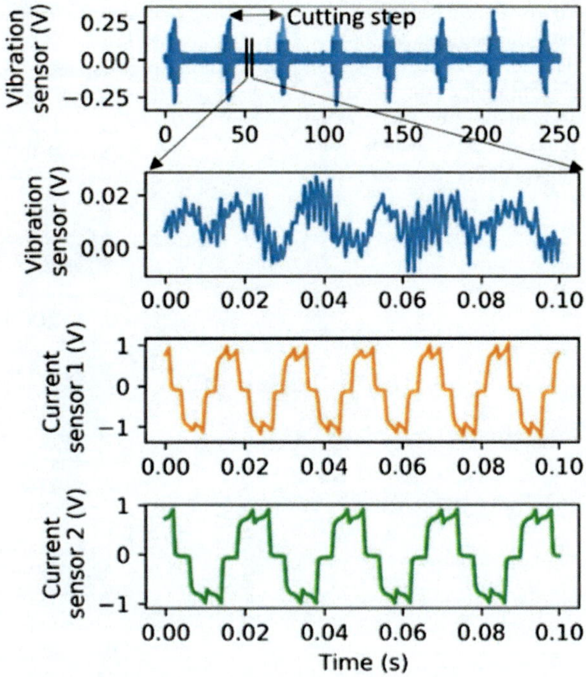

Fig. 4 Wafer cutting data. From top to bottom: Vibration data in full, zoomed-in vibration data, and zoomed-in motor current data for both sensors. Lumps in the vibration data are caused by the blade's sharp accelerations and decelerations at the beginning and end of a cutting step

neural networks, good results were also achieved on image processing tasks such as segmentation [20] or restoration [21, 22]. The parameters for the RNN are based on the results of a comprehensive study of encoder–decoder performances in automated translation [23]. While these results are about text translation and might not readily apply in our anomaly detection situation, we decided to keep them as guidelines due to the time cost of carrying out a similar study by ourselves.

Based on this study, we built our encoder–decoder model as described in Sect. 3.2. A few differences should, however, be noted between the network architectures used. Given that, our primary target is to keep the system as simple as possible. We stripped the RNN from complex optimizations such as attention mechanism and beam search, and we also used mono-directional RNNs instead of bidirectional ones. The other RNN parameters such as the number of time steps, cell input size, and learning rate were determined by grid search and are summarized in Table 1.

3.1 LSTM Cells

LSTM [13] cells are an improvement on classical recurrent cells that solve most of the issues inherent to RNNs. The LSTM cells rely on data control operations, called "gates," to move the information around in a way that makes sense for the neural network. The equations governing LSTMs operation are as follows:

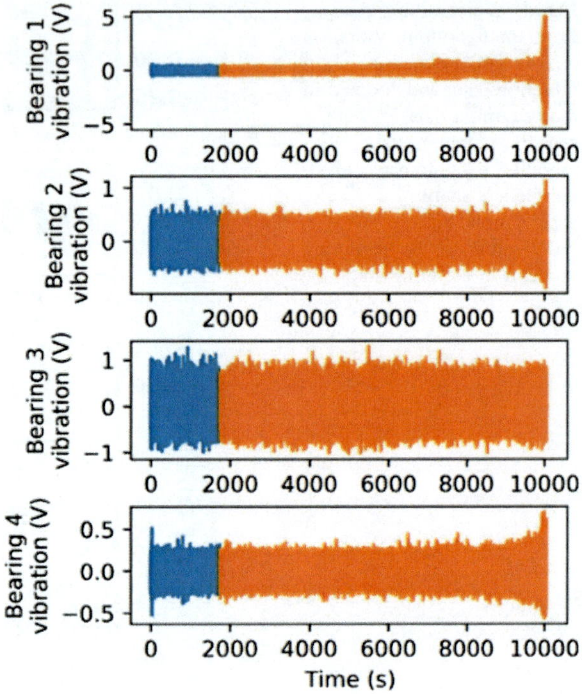

Fig. 5 Bearing vibration data, training data are in blue (left side of the figure), testing data in orange (right side). The failure of bearing 1 at the end of the experiment is easily noticeable

Table 1 Encoder–decoder parameters

Parameter	Value (Wafer cutting dataset)	Value (IMS dataset)
Encoder layers	2	2
Decoder layers	2	2
Number of time steps	10	5
Input vector size	500	100
Output vector size	500	100
Dropout keep rate	0.6	0.6
Batch size	25	25
Learning rate	9e-4	1e-5

$$\text{in}_f = W_f X_t + U_f Y_{t-1} + b_f \tag{1}$$

$$f_t = \sigma\left(\text{in}_f\right) \tag{2}$$

$$\text{in}_i = W_i X_t + U_i Y_{t-1} + b_i \tag{3}$$

$$i_t = \sigma(\text{in}_i) \tag{4}$$

$$\text{in}_z = W_z X_t + U_z Y_{t-1} + b_z \tag{5}$$

$$z_t = g(\text{in}_z) \tag{6}$$

$$\text{in}_o = W_o X_t + U_o Y_{t-1} + b_o \tag{7}$$

$$o_t = \sigma(\text{in}_o) \tag{8}$$

$$C_t = C_{t-1} f_t + z_t i_t \tag{9}$$

$$Y_t = h(C_t) o_t \tag{10}$$

With the following notations, noting that dimensions h and d are, respectively, the number of hidden units and the input dimension:

- $X_t \in \mathbb{R}^d$: LSTM unit input vector
- $C_t \in \mathbb{R}^h$: Cell state vector
- $Y_t \in \mathbb{R}^h$: LSTM unit output vector
- $W \in \mathbb{R}^{h \times d}$ and $U \in \mathbb{R}^{h \times h}$: weight matrices for the input and recurrent inputs
- $b \in \mathbb{R}^h$: bias vector
- g and h are, respectively, the input activation function and the output activation function (usually taken as tan$h$)

A schematic of a LSTM cell that summarizes these equations is shown Fig. 6.

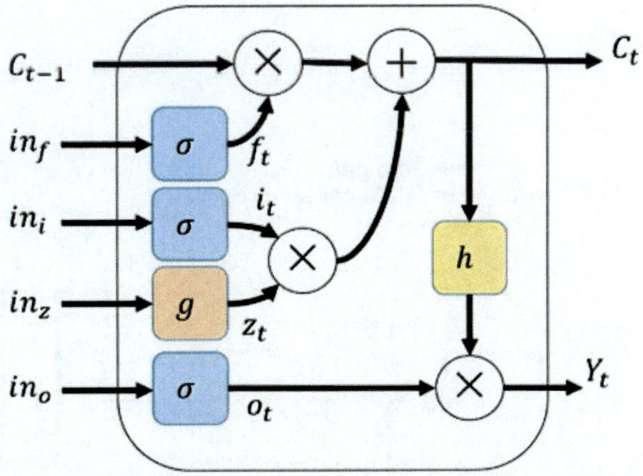

Fig. 6 LSTM cell schematic showing the various gates and the constant error carousel, also called cell state

3.2 Encoder–Decoder Architecture

Both the encoder and the decoder have two layers of LSTM cells, only the cell states are communicated between the encoder and the decoder. The decoder output is used as input for two fully connected layers that compute the final result. Individual LSTM cell outputs in the decoder are fed back into the next time step's input. A schematic of the architecture used is shown in Fig. 7. Dropout [24] is applied to the input and to the LSTM cells with a keep rate of 0.6.

The objective is to train the RNN to generate new healthy sensor values from the sensor data collected in healthy state. The main advantage of the encoder–decoder architecture when compared with others is that contrary to the architectures used in previous studies, we can train the RNN from raw data only, dispensing with the need for advanced preprocessing such as FFT, filtering, and wavelet transform. Once the training is over, in order to evaluate the machine state, we use the unseen machine sensor values as input and compare the RNN output to the sensor values we know are true. If the unseen data are similar to the data used during training, the RNN should be able to predict the sensor values reliably. On the other hand, if the unseen data differ from the training data, meaning the machine is no longer in a healthy state, the RNN should predict wrong values. By checking the quality of the predictions against the sensor values, we can therefore detect which state the system is in. This is a technique known as auto-associative neural network [25]. Note that the predictions do not have to be very precise; we just need them to yield a relatively constant error on healthy data and a noticeably different error on anomalous data. The choice of the input–output couple depends on the dataset used and is detailed hereafter.

When processing wafer cutting data, the encoder is fed with slices of the vibration data, and each input vector is made of 500 samples. Ten consecutive input vectors constitute a complete training example, thus, totaling 5000 data points

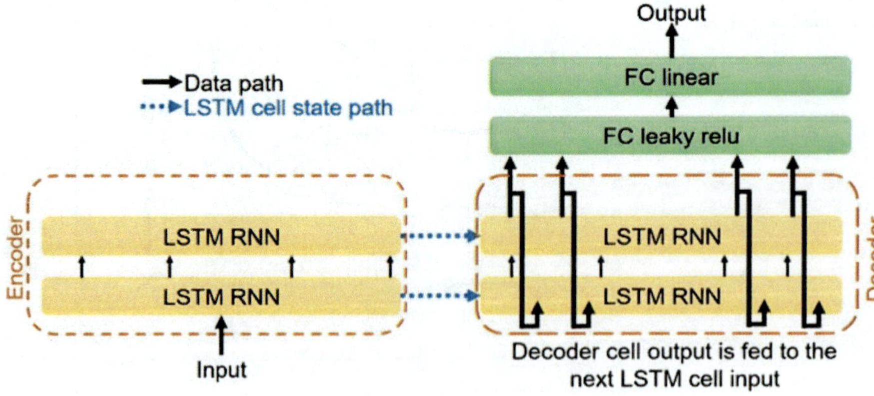

Fig. 7 Structure of the encoder–decoder network used for the experiments

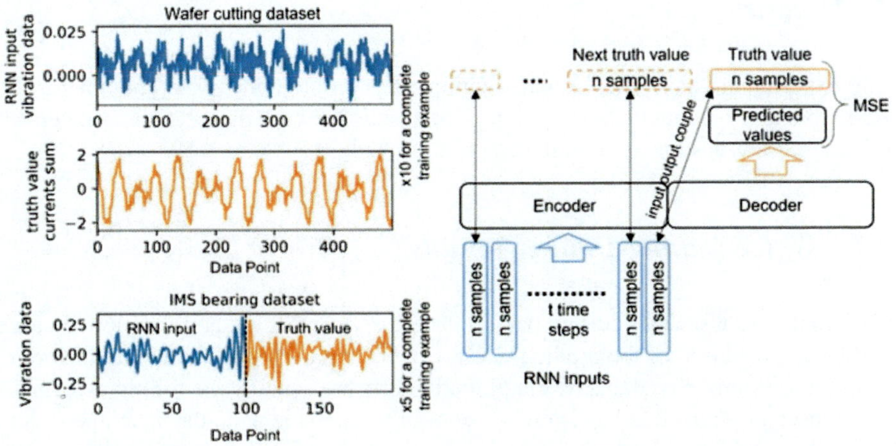

Fig. 8 Couples input–output for both datasets, RNN inputs (blue) are the value used as input and truth values (orange) are the values the RNN will try to predict. On the wafer cutting dataset, the values to predict have the same timestamp than the input. On the IMS bearing dataset, values to predict are contiguous to the input

representing 0.25 s of real-time. When using the IMS bearing data, the input vector contains 100 samples and a training example contains 5 input vectors corresponding to 500 data points or 0.025 s of measured data. A single input vector for each dataset is shown in blue in Fig. 8. In both cases, the RNN is trained using Adam solver with batches of 25 examples at a time, and the dataset is randomly shuffled at the start of every epoch. Gradient clipping [26] is applied during the training with a maximum gradient norm of 5. The loss function used is the mean squared error (MSE) between the output of the RNN, and the truth value obtained from the sensors reading is shown in (11).

$$\text{loss} = \sim \frac{1}{N}\sum_{1}^{N}(\text{output} - \text{truth value})^2 \quad (11)$$

Regarding the truth value, on the wafer cutting dataset, we tried empirically various combinations of the sensors. It turns out that by summing both motor current sensors' values and using the result as the truth for RNN predictions, we can get much better results than with complex filtering and/or feature engineering. This shows that label engineering might be a way to improve performance when changing the truth values is possible. For the IMS dataset experiments, the truth values are simply the next 100 values after the 100 input samples since there is no meaningful way to combine several sensors' data. A summary of the input–output situation is shown in Fig. 8.

4 Results

The experiments have shown that training for a single epoch was enough for the RNN to learn to generate healthy data. The results for both datasets after one epoch of training are described hereafter.

4.1 Wafer Cutting Dataset Results

The results for the wafer cutting dataset can be seen in Fig. 9, and the plotted curve shows the evolution of the prediction error over time during the machine operation. In order to improve readability, the plotted values are actually the rolling average of ten consecutive error values; this gives us an assessment of the machine health every 2.5 s which is enough for real-time monitoring in our situation. Some data points in the healthy and damaged zones were omitted to reduce the size of the figure and to put more emphasis on the part where the error increases. As said earlier, we do not need extremely precise predictions but only constant error on the healthy part and a higher error when the system is breaking down. The error in healthy state hovers around 0.6 to 0.8 and increases steadily to reach a plateau at around 1.6 when the blade is nearing the end of its lifetime. In real-time, this error increase occurs in about 3 days and the plateau lasts for roughly 2 days, proving that it is possible to detect the blade deterioration long before its actual failure thanks to the RNN predictions. Why the error stops increasing after some point is still unclear and needs more investigation.

4.2 IMS Bearings Dataset Results

Figure 10 shows the results obtained on the bearing predictions for all four channels. Three main areas are delimited:

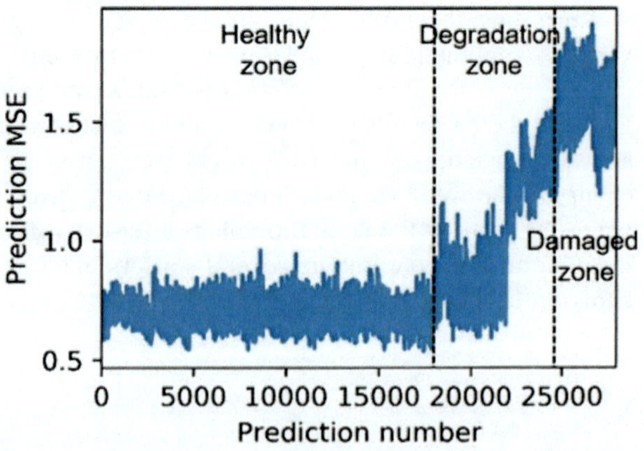

Fig. 9 Wafer cutting dataset prediction error results, and the error increases steadily with the blade degradation until a plateau. Blade breakdown occurred shortly after the rightmost point

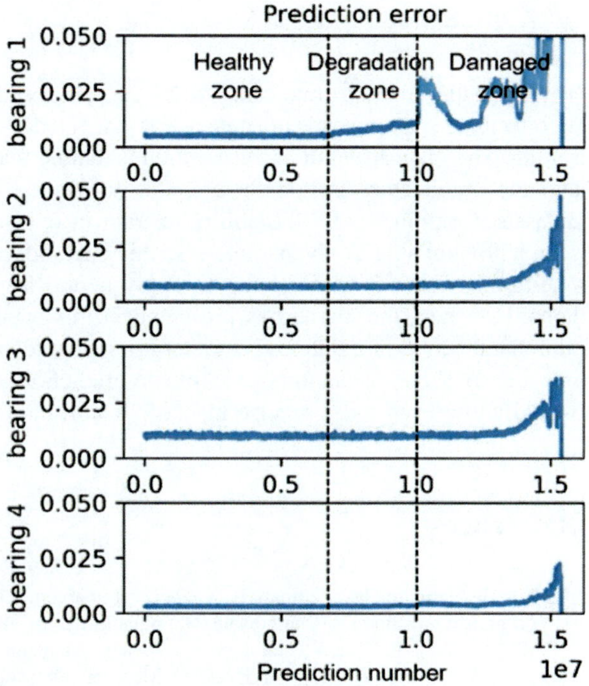

Fig. 10 IMS bearings test set results. Three working zones can be observed; the critical zone regarding machine health monitoring is the central one, where there is still time to intervene before the bearing failure

- The healthy zone where no bearing is showing any signs of fatigue. This is the normal state of the system.
- The degradation zone where the prediction error slowly increases in bearing 1 data while the other channels keep a constant error. This is the important zone where machine health monitoring actually provides useful information, showing that bearing 1 is no longer operating in its nominal state while the others are still fine. Steps should be taken to carry out an inspection and probably replace the bearing.
- The damaged zone where the error increases rapidly and bearing 1 is quickly failing. In this zone, the machine health monitoring system is not really providing anything useful since a human operator can see that there is something wrong simply by watching the sensor data with the naked eye.

In the damaged zone, the other bearings' data prediction errors also increase. This is due to the additional stress placed on the remaining healthy bearings when bearing 1 starts failing and no longer helps to support the load. This further shows the importance of being able to carry out maintenance operations early in the degradation zone in order to avoid damaging the other bearings by having them operating outside of their specification range in case of an unexpected failure.

It is also important to note that our results are qualitatively similar to those obtained in previous studies using feature extraction [7, 27], giving us confidence in the validity of our method.

5 Conclusion

These results demonstrate that with the proposed recurrent encoder–decoder architecture, it is possible to detect component degradation in a noisy industrial setting. The main strength of this method is that no feature engineering is necessary, and the architecture is nonetheless able to work reliably on two very different datasets. Combined with the ability to train in only one epoch, enabling fast grid search for suitable RNN parameters, this method can save a lot of design and fine-tuning time for those trying to apply neural network solutions to their problems. However, the method as presented in this paper can only give a qualitative hint that the system monitored is undergoing degradation. Further works are needed in order to obtain a remaining useful life prediction as the output of the RNN that would allow even more precise machine health monitoring.

References

1. Luo J, Namburu M, Pattipati K, Qiao L, Kawamoto M, Chigusa S (2003) Model-based prognostic techniques [maintenance applications]. In: Proceedings of AUTOTESTCON. IEEE systems readiness technology conference, Anaheim, CA, USA, pp 330–340
2. Yang Z, Djurdjanovic D, Ni J (2008) Maintenance scheduling in manufacturing systems based on predicted machine degradation. J Intell Manuf 19:87–98. https://doi.org/10.1007/s10845-007-0047-3
3. Cassady CR, Kutanoglu E (2003) Minimizing job tardiness using integrated preventive maintenance planning and production scheduling. IIE Trans 35:503–513. https://doi.org/10.1080/07408170390187951
4. Gurel S, Akturk MS (2008) Scheduling preventive maintenance on a single CNC machine. Int J Prod Res 46(24):6797–6821. https://doi.org/10.1080/00207540701487833
5. Martínez-Rego D, Fontenla-Romero O, Alonso-Betanzos A (2011) Power wind mill fault detection via one-class ν-SVM vibration signal analysis. In: International Joint Conference on Neural Networks, San Jose, CA, pp 511–518
6. Liu Z, Mei W, Zeng X, Yang C, Zhou X (2017) Remaining useful life estimation of insulated gate biploar transistors (IGBTs) based on a novel volterra k-nearest neighbor optimally pruned extreme learning machine (VKOPP) model using degradation data. Sensors 17 (11):2524. https://doi.org/10.3390/s17112524
7. Ben Ali J, Fnaiech N, Saidi L, Chebel-Morello B, Fnaiech F (2015) Application of empirical mode decomposition and artificial neural network for automatic bearing fault diagnosis based on vibration signals. Appl Acoust 89:16–27
8. Qiao H, Wang T, Wang P, Qiao S, Zhang L (2018) A time-distributed spatiotemporal feature learning method for machine health monitoring with multi-sensor time series. Sensors 18 (9):2932. https://doi.org/10.3390/s18092932
9. Guo L, Li N, Jia F, Lei Y, Lei J (2017) A recurrent neural network based health indicator for remaining useful life prediction of bearings. Neurocomputing 240:98–109. https://doi.org/10.1016/j.neucom
10. Liu X, Wu J, Zhou Z (2009) Exploratory undersampling for class-imbalance learning. IEEE Trans Syst Man Cybern Part B (Cybernetics) 39(2):539–550. https://doi.org/10.1109/TSMCB.2008.2007853

11. Seiffert C, Khoshgoftaar TM, Van Hulse J, Napolitano A (2010) RUSBoost: a hybrid approach to alleviating class imbalance. IEEE Trans Syst Man Cybern Part A Syst Hum 40 (1):185–197. https://doi.org/10.1109/TSMCA.2009.2029559
12. Khan SH, Hayat M, Bennamoun M, Sohel FA, Togneri R (2018) Cost-sensitive learning of deep feature representations from imbalanced data .IEEE Trans Neural Netw Learn Syst 29 (8):3573–3587. https://doi.org/10.1109/TNNLS.2017.2732482
13. Hochreiter S, Schmidhuber J (1997) Long short-term memory. Neural Comput 9(8):1735–1780
14. Lee J, Qiu H, Yu G, Lin J, Rexnord Technical Services (2007) IMS, University of Cincinnati. "Bearing Data Set". NASA Ames Prognostics Data Repository (http://ti.arc.nasa.gov/project/prognostic-data-repository), NASA Ames Research Center, Moffett Field, CA
15. Hai Q, Jay L, Jing L, Gang Y (2007) Wavelet filter-based weak signature detection method and its application on rolling element bearing prognostics. J Sound Vibr 289(4):1066–1090
16. Tobon-Mejia DA, Medjaher K, Zerhouni N, Tripot G (2010) A mixture of Gaussians Hidden Markov Model for failure diagnostic and prognostic. In: IEEE international conference on automation science and engineering, Toronto, ON, pp 338–343. https://doi.org/10.1109/COASE.2010.5584759
17. Ilya S, Oriol V, Quoc VL (2014) Sequence to sequence learning with neural networks. Proc NIPS 27:3104–3112
18. Thang L, Hieu P, Manning CD (2015) Effective approaches to attention-based neural machine translation. Proc EMNLP, 1412–1421
19. Lu L, Zhang X, Renais S (2016) On training the recurrent neural network encoder-decoder for large vocabulary end-to-end speech recognition. In: IEEE International conference on acoustics, speech and signal processing (ICASSP), Shanghai, pp 5060–5064. https://doi.org/10.1109/ICASSP.2016.7472641
20. Badrinarayanan V, Kendall A, Cipolla R (2017) SegNet: a deep convolutional encoder-decoder architecture for image segmentation. IEEE Trans Pattern Anal Mach Intell 39(12):2481–2495. https://doi.org/10.1109/TPAMI.2016.2644615
21. Xiaojiao M, Chunhua S, Yu-Bin Y (2016) Image restoration using very deep convolutional encoder-decoder networks with symmetric skip connections. Proc NIPS 29:2802–2810
22. Chen H et al (2017) Low-Dose CT With a Residual Encoder-Decoder Convolutional Neural Network. IEEE Trans Med Imaging 36(12):2524–2535. https://doi.org/10.1109/TMI.2017.2715284
23. Britz D, Goldie A, Luong MT, Le Q (2017) Massive exploration of neural machine translation architectures. Preprint arXiv:1703.03906
24. Srivastava M, Hinton G, Krizhevsky A, Sutskever LY, Salakhutdinov R (2014) Dropout: a simple way to prevent neural networks from overfitting. J Mach Learn Res 15(1):1929–58
25. Japkowicz N (2001) Supervised versus unsupervised binary-learning by feedforward neural networks. Machine Learning 42(1–2):97–122
26. Pascanu R, Mikolov T, Bengio Y (2013) On the difficulty of training recurrent neural networks. In: International conference on machine learning, pp 1310–1318
27. Mao W, Tian S, Fan J, Liang X, Safian A (2020) Online detection of bearing incipient fault with semi-supervised architecture and deep feature representation. J Manuf Syst 55:179–198

A Sizing System Using Anthropometric Measurements for Headgear

W. H. A. C. Wijerathna, M. M. I. D. Manthilake, and H. K. G. Punchihewa

1 Introduction

Most of the consumer products such as helmets and sunglasses are not compatible with the Asian populations since they are designed for the Caucasian population using their anthropometric data [1]. It is well known that there is a wide variation between the Asian and European countries in terms of anthropometric measurements, and it is the reason for the incompatibility of imported products in terms of fit and usability [2]. Unfortunately, all the countries, especially the developing countries, do not manufacture fitting products due to the lack of body size data and unawareness of the importance of ergonomics principles [2]. Therefore, attention on developing products for fit and usability forms an important area for research and development.

Use of inappropriate or ill-fitting safety gear such as helmets poses severe danger to the wearers [3], and this aspect needs special emphasis in the design and manufacturing arena. A survey conducted among international cricket players reveals that it is difficult for the cricketers to find headgear that gives a proper fit [4]. The level of protection that a helmet provides depends on the fit of the helmets and the correct wearing position at the moment of impact [3]. Wearing an under-sized or over-sized helmet may result in poor helmet fit and incorrectly wearing the helmet could lead to tilting of the helmet forward or backward increasing the possibility of injury during impact [4]. Research suggests that ill-fitting helmets or improper use of helmets cause injury when the impact energy is instantly transferred to the head and neck in situations of impact [5, 6]. Unfortunately, most of the users are unaware about the correct positioning of the helmet on the head, and they even tighten the chin strap to a level of discomfort [6] as a measure to accomplish acceptable fit. This may be because there is currently only a single-criterion sizing system based

W. H. A. C. Wijerathna · M. M. I. D. Manthilake · H. K. G. Punchihewa (✉)
Department of Mechanical Engineering, University of Moratuwa, Moratuwa, Sri Lanka
e-mail: himank@uom.lk

R. K. Agarwal (ed.), *Recent Advances in Manufacturing Engineering and Processes*, Lecture Notes in Mechanical Engineering,
https://doi.org/10.1007/978-981-16-3934-0_14

on the head circumference (HC) as shown in Table 1 for cricket headgear. In addition, both men and women that play cricket even at the international level are compelled to use the available sizes that have been established disregarding the gender differences in anthropometry [4] and any other variations in head shape among people [2, 7]. Female cricketers indicate that they find it difficult to select suitable headgear that provides good fit due to their small head sizes [4], and hence, they tend to increase the number of inner liners of the helmet to improve fit. Due to these modifications made by the female players, the thickness of the inner impact absorption padding increases, potentially causing the helmet to shift upwards relative to the head. This could lead to poor vision through the peak and the faceguard of the helmets. It could also increase the weight of the helmet and heat retention within the helmet. International cricket players further reveal the safety issues associated with the aforesaid shortcomings [4] signifying the need for an improved sizing and grading system for helmets/headgear.

To determine a suitable headgear for players, there should be a proper sizing system that can be easily understood and used in design and manufacture. At the moment, only the HC is used to determine the headgear size [8] as shown in Table 1 and in the present sizing system, only six sizes are available, i.e. small boy, boy, youth, small men, men and large men. It appears that the HC alone is not sufficient for a fitting helmet based on the results of previous studies.

Accomplishing perfect fit has always been a challenge for the product designers [9], and comprehensive anthropometric information is needed to ensure good fit [10, 11]. Developing headgear according to the head shape will also lead to reduced weight and ensure the stability of the helmet on the wearers [12]. According to the literature, the major strategies for ‘fitting design’ are adjustability in design, population grouping and customization [9]. However, helmet shells are hard, and it may be difficult to make them adjustable or flexible. Therefore, population grouping and customisation become the key approaches to design fitting helmets.

Considering anthropometry when designing consumer products ensures fit and usability [1]. In addition, research suggests that using anthropometric data in the conceptual design stage may minimize the size and shape alterations needed later [13] and also reduce the risk of injuries both in the short- and long-term [14, 15]. Similarly, to obtain a technically good fit in helmets, it is important to understand the shape of head as well (ref). Therefore, both sizing and grading become important criteria to develop helmets to provide fit to the target population. Sizing is

Table 1 The current cricket helmet/headgear size chart [8]

HC (cm)	Helmet/headgear size
51–52	Small boy
53–54	Boy
55–56	Youth
57–58	Small men
58–60	Men
61–62	Large men

the approach to develop a limited number of product sizes for a given population [16] whilst grading is the relationship between the sizes [16] that provides the method to change from one size to another. Furthermore, a new grading system that encompasses both male and female head anthropometry, different head shapes, sizes and other gender-based aspects such as having long hair and earrings become a necessity to help design headgear. Such approach could positively influence the fit, even though the number of sizes is kept fixed to a minimum. Therefore, the aim of the reported study was to develop a new sizing and grading system using anthropometric measurements for headgear, and the objectives were to determine a set of head sizes and a corresponding grading method to achieve better fit and usability.

2 Methodology

The Prisma method [17] was used to conduct a literature review to identify the key parameters for designing a helmet sizing system. Literature was accessed using databases such as Ergonomics abstracts, Sage, Science direct, Scopus and Elsevier based on a keyword search. The keywords for the selection of the literature were size and shape analysis of head, craniofacial anthropometry, size and grading of products, anthropometry for product design, ergonomics and design. After the title-based search, if any of the abstracts contained information related to craniofacial anthropometry and ergonomics design of helmets, such literature were selected for further reading.

Anthropometric landmarks are key sites found on the human body [10, 18]. A standardised set of landmarks are used to take anthropometric measurements [10]. Table 2 and Fig. 1 show the accepted anthropometric landmarks for taking craniofacial measurements. In this study, head circumference, head length and head breadth were used to design the new sizing system since these key measurements are correlated with other head and face measurements [19]. For example, head length has good correlation with head sagittal arc and head breadth has a high correlation with bitragion arc [19]. According to the ISO 7250–1:2008, a measuring tape can be used to measure the head circumference and the spreading calliper can be used to measure head length and head breadth [11].

In order to develop the sizing system, anthropometric measurements were gathered using different sources [2, 23–26]. Using the set of equations in Table 3,

Table 2 Anthropometric landmarks for taking craniofacial measurements [20, 21]

Measurement	Landmarks	
	Beginning	End
Head circumference (HC)	Occiput	Occiput
Head length (HL)	Glabella	Occiput
Head breadth (HB)	Left euryon	Right euryon

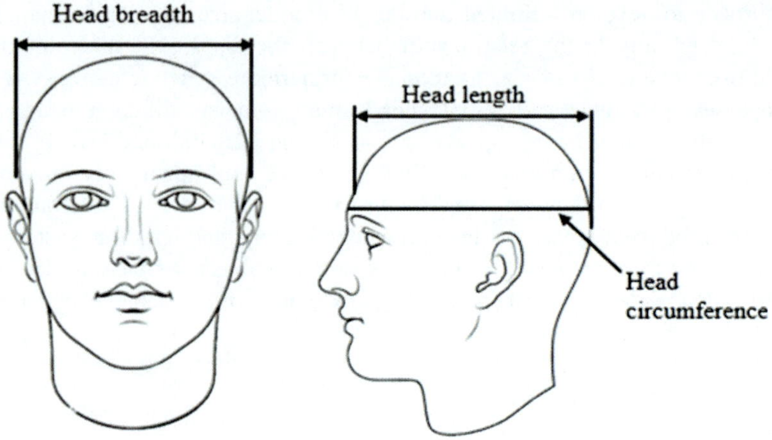

Fig. 1 Craniofacial measurements [22]

Table 3 Formula for combining groups [27]

	Group 1	Group 2	Combined groups
Sample size	N1	N2	N1 + N2
Mean	M1	M2	$\frac{N_1M_1+N_2M_2}{N_1+N_2}$
SD	SD1	SD2	$\sqrt{\frac{(N_1-1)SD_1+(N_2-1)SD_2+\frac{N_1N_2}{N_1+N_2}\left(M_1^2+M_2^2-2M_1M_2\right)}{N_1+N_2-1}}$

the mean and the standard deviation of HC, HB and HL were calculated for the Asian and Caucasian populations above nine years old.

The 5th and 95th percentiles of the head sizes were obtained using the calculated mean and the standard deviation values for the selected populations. All these measurements were then used to develop a regression line. The regression line was created using Microsoft Excel® and the upper bound and lower bound were determined by creating parallel lines to the regression line such that all the points were between the upper boundary line and lower boundary line. Then, the sizing system was established to include the 5th percentile of the Asian nine-year-old females to the 95th percentile of the Caucasian adults in terms of the HC, HL and HB. The headgear sizes were categorised using the same step (i.e. 2 cm) based on the HC that is being used in the current helmet sizing system.

The ratio between the head breadth (HB) and head length (HL) as a percentage is called the Cephalic index, and it can be used as a method of describing the head shape [20]. Based on the Cephalic index, three basic head shapes are identified, namely, dolichocephalic, mesocephalic and brachycephalic [20]. Figure 2 shows head shapes according to the Cephalic index. These shapes need to be accommodated in a grading system. Therefore, the inner padding thicknesses for the helmets

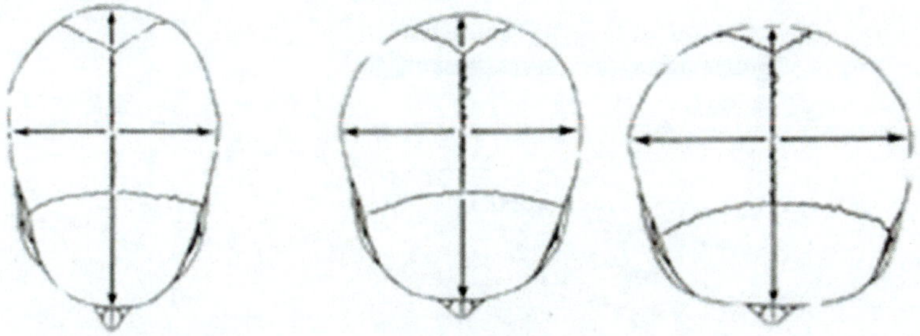

Fig. 2 Head shapes according to the Cephalic index [20, 21, 28]

were determined considering the head shapes within the established helmet sizes. As such, foam thickness for padding for dolichocephalic, mesocephalic and brachycephalic head shapes were established for each helmet size.

$$\text{Cephalic} \sim \text{index} = \frac{\text{Head} \sim \text{breadth}}{\text{Head} \sim \text{length}} \times 100\%$$

3 Results

The literature review was used to identify key anthropometric parameters for designing the helmet sizing system. Those are: head circumference (HC), head length (HL) and head breadth (HB) and these key parameters show correlation with other craniofacial measurements.

Figure 3 shows the HB and HL values against the HC for different populations. In this graph, the points in orange colour represent the HB and those in blue colour represent the HL. The corresponding continuous lines represent the regression lines for the two attributes and the two dashed lines drawn parallel to the regression lines in each of the graphs represent the upper and lower bounds for the two parameters.

A wearable product should include all the individuals who are within the upper and lower bounds. The vertical red line shows the smallest helmet size according to the existing helmet size chart. Many individuals in the Asian population are on the left side of the red line and many of the Caucasians are on the right side of the red line. This essentially means that those who have head circumferences smaller than 51 cm have not been included when developing the helmets currently being used. Therefore, new sizes were added to the lower end of HC in the current size chart.

Thus, the two smaller sizes, i.e. XS and XXS, are proposed. The new sizing chart is presented in Table 4 with corresponding values for HB and HL.

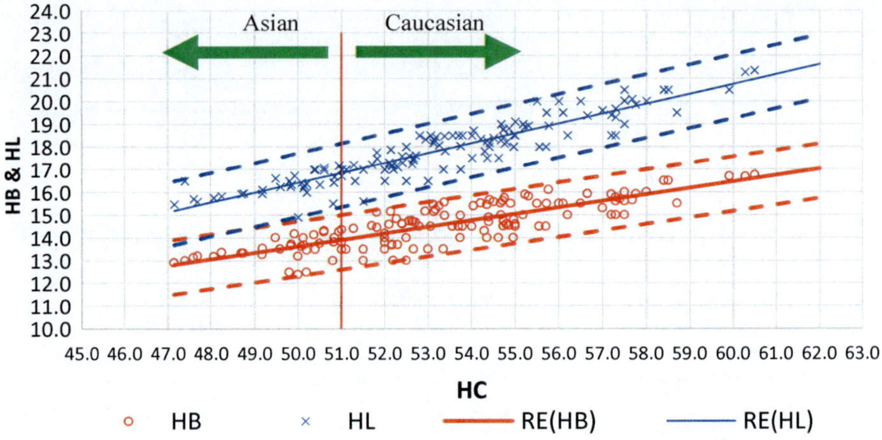

Fig. 3 Head breadth (HB) and head length (HL) against head circumference (HC)

Table 4 Developed size chart for the helmet shell

Helmet size	HC (cm)	HB (cm)	HL (cm)
XXS	47–48	11–15	13–17
XS	49–50	12–15	14–18
S	51–52	13–16	15–19
M	53–54	13–16	16–20
L	55–56	14–17	17–21
XL	57–58	14–17	18–22
XXL	59–60	15–18	19–23
XXXL	61–62	15–18	20–23

Table 5 Types and thicknesses of the inner protective padding

D foams (mm)	M foams (mm)	B foams (mm)
8	8	8
10	10	10
12	12	12

Table 5 shows the three types of foams and their thicknesses selected to accommodate the different head shapes; 8, 10 and 12 mm are the standard thicknesses currently being used. For the grading system, three foams are proposed, i.e. D foams, M foams and B foams. They accommodate the three different head shapes, dolichocephalic, mesocephalic and brachycephalic, respectively. Table 6 shows the developed grading chart to help identify the suitable inner protective padding thicknesses to suit the different head shapes. Table 6 is developed by combining Tables 4 and 5.

Table 6 Grading chart for headgears

		Thickness of the inner protective padding (mm)																
Helmet size	HC (cm)	46–47	47–48	48–49	49–50	50–51	51–52	52–53	53–54	54–55	55–56	56–57	57–58	58–59	59–60	60–61	61–62	62–63
XXS	47–48	12	10	8														
XS	49–50			12	10	8												
S	51–52					12	10	8										
M	53–54							12	10	8								
L	55–56									12	10	8						
XL	57–58											12	10	8				
XXL	59–60													12	10	8		
XXXL	61–62															12	10	8

4 Discussion

In the present sizing system, there are only six helmet sizes to accommodate the head circumference ranging from 51 to 62 cm, and it appears that the 5th percentile head circumference, which is 50 cm for the nine-year-old Asian females [2, 25] does not fall within this range. The literature also reveals that it is difficult to find a headgear that gives a proper fit for women cricket players [4]. The developed size chart introduces two additional sizes, and it will help to find suitable headgear for children as well.

Three key measurements have mainly been used in this developed size chart: head length, head breadth and head circumference. A measuring tape can be used to measure the head circumference and the spreading calliper can be used to measure head length and head breadth [11]. Using the head circumference, suitable helmet shells can be identified. After that, to calculate the Cephalic index, head breadth and head length need to be used. Using the Cephalic index, the head shape is identified, and this can lead to the selection of the suitable inner protective padding thickness for a given player. The developed size and grading charts (Tables 4 and 5) can be used in tandem to decide on the suitable headgear size and inner protective padding thickness for a given head size and shape. In order to ease the burden of referring to two separate tables, Table 6 has been developed combining Tables 4 and 5. This novel sizing and grading chart for cricket helmets can be used to conveniently select fitting headgear for both male and female players. As an example, if a person can wear a particular helmet size with 8 mm thickness inner protective padding, same person can wear the next plus size with a 12 mm thickness inner protective padding. If a particular person can wear an XXS shell with D foams and 8 mm thickness inner protective padding, the same person should be able to wear an XS shell with D foams and 12 mm thickness inner protective padding. Thus, the wearer could select the shell and inner protective padding that gives the best comfortable fit. Same relation can be applied for M foams and B foams and other shell sizes as well. Further, this grading chart can be used for all three head shapes [20] discussed in the literature.

Only the anthropometry of Asian cricketers (e.g. Indian and Sri Lankan) and Caucasians (e.g. Australian and British) were considered in this study to develop the sizing and grading system. Thus, it is required to use anthropometric measurements of other possibly different segments of the world population such as Africans and Chinese to ascertain adequacy of the developed size grading system. Further, the currently available inner protective padding thicknesses have been directly used to develop the grading system. However, further research needs to be carried out in order to identify the minimum and maximum thicknesses for the inner protective paddings for sufficient impact energy absorption and to have a better heat dissipation ability which was overlooked in the current study.

5 Conclusion

Three key head anthropometric measurements, i.e. head length, head breadth and head circumference, were used in the development of a novel size chart for cricket helmets. This novel sizing and grading chart for cricket helmets can be used to select fitting headgear for both male and female players of Asian and Caucasian origin over nine years old. It has eight different sizes as opposed to six sizes in the current sizing chart. Cephalic index is used to categorise the head shapes, and this was used to select the inner protective padding thicknesses to accommodate the different head shapes. The developed sizing and grading charts can be used in tandem to decide the suitable headgear size and inner protective padding thickness for given head size and shape. Future research is needed to ascertain the generalizability of the sizing and grading system for a wider population and identify the minimum and maximum thicknesses of paddings for sufficient impact energy absorption and better heat dissipation.

Acknowledgements The authors would like to acknowledge SRC (SRC/ST/2021/01) for allocating funds to conduct this research.

References

1. Ball R (2009) 3-D design tools from the Sizechina Project. Ergon Des 17(3):8–13
2. Abeysekera JD, Shahnavaz H (1987) Body size data of Sri Lankan workers and their variability with other populations in the world: its impact on the use of imported goods. J Hum Ergol (Tokyo) 16(2):193–208

3. Perret-Ellena T, Subic A, Pang TY, Mustafa H (2014) The helmet fit index a method for the computational analysis of fit between human head shapes and bicycle helmets. In: icSPORTS 2014-proceedings of the 2nd international congress on sports sciences research and technology support, no October, pp 145–153
4. Wijerathna WHAC, Manthilake MMID, Punchihewa HKG (2020) Shortcomings and user requirements associated with cricket helmets currently in use: A study on features for safety and usability. Int J Ind Ergon 80(June)
5. James Donald Jones J (2011) Custom fit helmet and its method of making US20110203038A1
6. Connor ATA et al (2016) Current standards for sports and automotive helmets : a review. Ref Ares 3151745:1–42
7. Tierney RT et al (2008) Sex differences in head acceleration during heading while wearing soccer headgear. J Athl Train 43(6):578–584
8. Allroundercricket.com, "Helmet sizes" [Online]. Available: https://www.allroundercricket.com/guides/cricket_helmets/size_guide/. Accessed: 12-Feb-2020
9. Liu H, Li Z, Zheng L (2008) Rapid preliminary helmet shell design based on three-dimensional anthropometric head data. J Eng Des 19(1):45–54
10. Roebuck JA (1995) Anthropometric methods: designing to fit the human body. Human factos and Ergonomics Society
11. Pheasant S (1988) Bodyspace: anthropometry, ergonomics and design. Taylor & Francis Ltd.
12. Ball R (2011) Human factors in protective headgear design, no. January 2011, pp 301–314
13. Robinette KM (1993) Fit testing as a helmet development tool. In: Human factors and Ergonomics society 37th annual meeting, pp 69–73
14. Punchihewa HKG, Gyi DE (2016) Reducing work-related Musculoskeletal Disorders (MSDs) through design: views of ergonomics and design practitioners. Work 53(1):127–142
15. Punchihewa HKG, Gyi DE (2009) Development of a QFD based collaborative design approach to reduce work-related musculoskeletal disorders (MSDs). Des Princ Pract An Int J Ann Rev 3(6):209–224
16. Luximon A, Zhang Y, Luximon Y, Xiao M (2012) Sizing and grading for wearable products. CAD Comput Aided Des 44(1):77–84
17. Moher D et al (2009) Preferred reporting items for systematic reviews and meta-analyses: the PRISMA statement. PLoS Med 6(7):1–6
18. Guo F, Wang L, Dong D (2009) Human head 3D dimensions measurement for the design of helmets. Lect Notes Comput Sci (including Subser Lect Notes Artif Intell Lect Notes Bioinform) 5620(LNCS):624–631
19. Ran L, Zhao C, Luo H, Zhang X, Liu T (2020) A study on the correlation of head and face data of chinese adults. Int Conf Appl Hum Factors Ergon 955:325–333
20. Hossain MG, Saw A, Alam R, Ohtsuki F, Kamarul T (2013) Multiple regression analysis of anthropometric measurements influencing the cephalic index of male Japanese university students. Singapore Med J 54(9):516–520
21. Croney J (1981) Anthropometry for designers: Revised edition. Van Nostrand Reinhold Company
22. Hand drawn human head in face [Online]. Available: https://www.alamy.com/hand-drawn-human-head-in-face-profile-and-top-views-flat-vector-isolated-on-white-background-the-drawings-can-be-used-independently-of-each-other-image212143006.html. Accessed: 14-Oct-2020
23. Norris B, Wilson JR (1995) Child data the handbook of child measurements and capabilities-data for design safety. Department of Trade and Industry
24. Peebles L, Norris B (1998) Adult data the handbook of adult anthropometric and strength measurements-data for design safety. Department of Trade and Industry
25. Chakrabarti D (1997) Indian anthropometric dimensions for ergonomics design practice. National Institute of Design

26. Nanayakkara D, Chandrasekera M (1998) Craniofacial anthropometry of Sri Lankans. Ceylon J Med Sci 41(1):7
27. Higgins JP, Green S (2008) Cochrane handbook for systematic reviews of interventions
28. "Head shapes." [Online]. Available: https://forums.skadi.net/threads/175307-Dolichocephalic-or-Brachycephalic. Accessed: 14-Oct-2020

A Methodology to Apply Process Mining in End-To-End Order Processing of Manufacturing Companies

G. Schuh, A. Gützlaff, S. Schmitz, C. Kuhn, and N. Klapper

1 Introduction

To compete in fast-paced environments, manufacturing companies describe their process performance (PP) in order to assess their competitiveness. PP measures processes' progress towards their objectives [1] and a process consists of numerous sub-processes and activities. Describing PP includes mapping as-is processes and measuring process performance indicators (PPI). Thus, PPI need to be unequivocally determinable [2].

Assessing the competitiveness of manufacturing companies raises the need to describe the PP of the entire end-to-end order processing process (ETEOPP) [3]. The ETEOPP includes all technical-operative core processes, reaching from sales processes and manufacturing processes to shipping processes and describes the sequence of operational processes transforming customer inquiries into saleable products [4]. Notwithstanding, as 96% of process optimisation projects are realized in manufacturing processes, most ETEOPP sub-processes are disregarded in PP descriptions. However, disregarded sub-processes make up 70% of the end-to-end process time. As a result, not taking the ETEOPP into consideration results in crucial non-transparencies for PP improvements [5]. Limitations are biased participants, large time consumption and limited abilities to capture process dynamics in paper-based techniques [6]. Further industry insights show that describing the ETEOPP is a significant problem. 62% of companies have only documented less than 25% of their processes and only 2% of companies have an overview of their complete process landscape [7].

G. Schuh · A. Gützlaff · S. Schmitz (✉) · C. Kuhn · N. Klapper
Laboratory for Machine Tools and Production Engineering (WZL), RWTH Aachen University, Campus-Boulevard 30, 52074 Aachen, Germany
e-mail: s.schmitz@wzl.rwth-aachen.de

R. K. Agarwal (ed.), *Recent Advances in Manufacturing Engineering and Processes*, Lecture Notes in Mechanical Engineering,
https://doi.org/10.1007/978-981-16-3934-0_15

Process mining (PM) can be applied to tackle deficits in process descriptions with a fact-based, objective and precise method. PM aims to discover, monitor and improve business processes using event data stored in event logs. However, PM has only been applied to single departments and partial processes with similar order types, respective order-IDs [8]. A three-phase framework is introduced in previous work to address PM in ETEOPP and shows the impact of data-based approaches on process analysis [3]. This paper provides an approach to merge multiple order types and calculates PPI as well as process models to expand the second phase of the framework. The remainder of this paper is structured as follows: Sect. 2 outlines the importance of PM for order processing. Section 3 presents the methodology for merging multiple event logs to apply PM across the ETEOPP. Section 4 validates the methodology using a dataset. In Sect. 5, the results of the paper are summarised and an outlook on further research is given.

2 Importance of Process Mining for Order Processing

Due to diverse order types, parallel or sequential activities in ETEOPP, process variances are often higher than assumed in manufacturing companies [9]. In the following, prerequisites for the application of PM in ETEOPP are outlined.

Process discovery as one type of PM algorithmically converts event log data into a process model [10] and quantifies indicators such as frequency, duration or throughput times. Regarding the ETEOPP, process models must display event data emerging from different departments of a company. However, event data of the ETEOPP are scattered across multiple information systems such as customer relationship management systems, enterprise resource planning systems and manufacturing execution systems [11]. Thus, data from multiple information systems must be defined in a data model and merged in an event log before PM techniques can be applied.

In the ETEOPP, order-IDs appearing in events can be categorised by different object types (OTs). Each OT characterises orders that are processed in partial business processes. For instance, customer orders (order-IDs of sales processes as one OT) could contain several articles represented by multiple manufacturing orders (order-IDs of manufacturing processes as a second OT). A customer order can be split and joined in various OTs throughout the ETEOPP. Resulting multiple order-IDs must be considered as process instances for evaluable results of PP descriptions across the ETEOPP [12].

The eXtensible Event Stream (XES) is the common format for event logs and PM applications but only represents one single OT [13]. A different format is required to represent multiple OTs for an ETEOPP. An object-centric event log (OCL) combines multiple OTs within a single data table [14]. In this paper, an OCL is a two-dimensional, column-structured table with multiple OTs (respective order-IDs), related activities and timestamps as data attributes [14]. This enables the tracing of orders with multiple order-IDs across processes. However, describing the PP requires transforming the OCL into an XES-structured data table to apply traditional PM algorithms.

In industry, widespread uncertainty exists regarding the suitability of available data for data-based analysis [15]. Thus, data requirements for data-based PP descriptions must be clearly defined. For PM applications in ETEOPP, it is assumed that, according to the first guiding principle [16], partial event logs are available in sufficient quality (i.e. without noise). Exception is syntactic data inhomogeneity, which results from merging multiple event logs of different information systems. Therefore, an application of PM must consider appropriate data preparation to improve the quality of resulting process models. Lastly, PPI that describe process efficiencies must be calculated for processes, traces and activities.

3 Methodology

The proposed methodology considers multiple event logs and varying OTs to describe the ETEOPP by mapping a process model and calculating PPI. The development is based on existing research to be rigour. First, [17] split up an OCL into an event log for every OT through flattening to apply established PM techniques. Second, describing PP requires mapping of as-is processes and calculating PPI. In [18] discovers a process model and enhanced it using separately calculated PPI before displaying results to the user. In [19], time-based PPI are calculated for the categories process, case and activity.

Figure 1 gives an overview of the six-step methodology. Section 3.1 describes the data tables (DTs) in detail as inputs of the methodology. First, the DT is combined into an OCL. Second, the OCL with multiple OTs is split into event logs for each OT. Third, event log traces are identified. In the remaining steps, the PPI and process model of the ETEOPP are calculated separately. Thus, fourth, PPI for the activity perspective are calculated. Fifth, PPI for the trace and process perspectives are calculated. Sixth, a process model for the smallest sub-instance OT of the ETEOPP is calculated. The outputs of the methodology are PPI for the

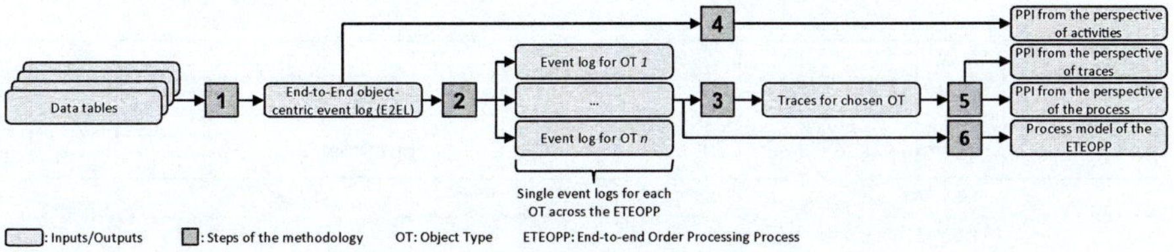

Fig. 1 Six-step methodology as well as their inputs and outputs

perspectives activity, trace and process as well as a process model of the ETEOPP to describe the PP of manufacturing companies.

3.1 Detailed Description of the DT as Inputs for the Methodology

Each DT is an extracted event log of a partial, department-specific process within a company's ETEOPP (e.g. sales, manufacturing, etc.). A DT is a two-dimensional, column-structured table with order-IDs as process instances as well as their related activities and timestamps as data attributes. The DT is comparable to the XES-Standard. The timestamps must record the start, the end and the planned end of the activity as well as the time when the order was received. These timestamps are necessary for calculating the PPI for the ETEOPP, which are elaborated in Sect. 3.2. The extraction and filtration of the DT from information systems are out of scope of the methodology.

3.2 Detailed Description of Step One to Three of the Methodology

In step one, the DT is merged to an OCL and extended to trace the ETEOPP from the viewpoint of every OT. To map the ETEOPP, related objects across all OTs need to be identified. Two objects across different OTs are related to each other if they occur in the same event within the OCL. The OCL is extended so that every time an object-related order-ID is treated within one event, related objects are complemented to the event. In this paper, the enriched OCL is called end-to-end OCL (E2EL). An example of the extension from OCL to E2EL is shown in Fig. 2. In the E2EL, the order numbers 990001 and 990,002 can be traced when their related shipping order was packaged in the third event. As a result, the ETEOPP, e.g. of an order number, can be mapped correctly, so that it also includes the packaging activity besides the initiation of an order.

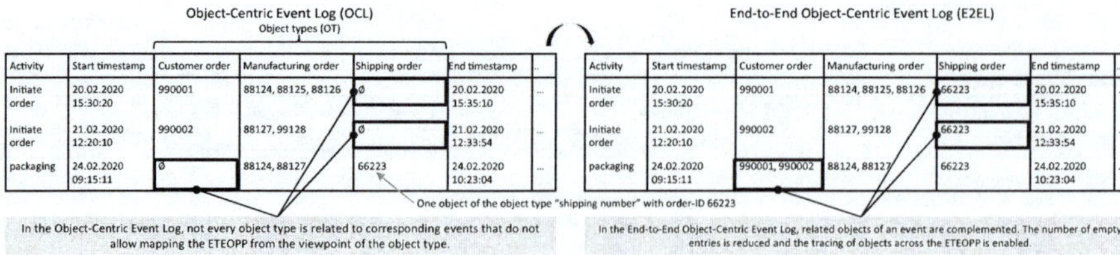

Fig. 2 Exemplary visualisation of the extension from OCL to E2EL

In step two, the E2EL is split into a DT for each OT. Each OT is selected as a case notion and the E2EL is flattened towards the selected OT. Flattening leads to three problems—divergence, convergence, and deficiency [17]. Divergence is the loss of ordering information leading to loops in the process model that do not exist in the real process. Section 3.4 addresses the divergence problem. Convergence is the replication of an event that is executed for multiple objects, falsifying the real number of events. Section 3.3 further deals with the convergence problem. Deficiency describes the disappearance of events, which do not include objects of the selected OT. The E2EL diminishes deficiency, as the number of OTs included in every event is increased. The outputs of the second step are DT for every OT of the entire ETEOPP.

The input of step three is the resulting DT from step two. The DT events are separated according to their objects to create corresponding traces. By doing so, all event attributes are kept such that no information is lost during this step. The output of step three is traces for every object of the event log. The existence and placement of step three are justified due to the following reasons: first, an event log for every OT is required as input for step six, such that the step cannot be merged with step two. Second, this step prepares the data while step five calculates PPI. Thus, separating both steps allows a better understanding and distinction of the steps of the methodology.

3.3 *Detailed Description of Step Four and Five of the Methodology*

In step four, the PPI from the activity perspective are calculated. Input for step four is the E2EL. The E2EL is not modified by flattening, so the activity PPI are not affected by convergence. The five PPI process time, time of response, deadline adherence, mean tardiness and process reliability are calculated based on previous works [20]. In this paper, the calculations for the process time and the deadline adherence are further elaborated. Equation (1) depicts the calculation $\mathrm{PPI}_{\mathrm{PT,a}}$ for process time PT of activity a. Therefore, the sum over all events E in the E2EL is taken. Each event is filtered for the inquired activity using the expression in Eq. (2). The process time for each event i is calculated by subtracting the start timestamp $\mathrm{TS_i}$ from the end timestamp $\mathrm{TE_i}$. Thus, the process time of an activity is the average duration of all the instances of that activity.

$$\mathrm{PPI}_{\mathrm{PT},a} = \frac{\sum_i^E \left(x_{i,a} \times (\mathrm{TE}_i - \mathrm{TS}_i)\right)}{\sum_i^E x_{i,a}} \text{ with } \mathrm{PPI}_{\mathrm{PT},a} \in [0, \infty)\forall \mathrm{a} \tag{1}$$

$$x_{i,a} = \begin{cases} 1; \text{Event } i \text{ includes activity } a \\ 0; \text{Event } i \text{ does not include acitivty } a \end{cases} \quad \forall i, a \tag{2}$$

Equation (3) shows the calculation of $PPI_{DA,a}$ for the deadline adherence DA of an activity based on Eq. (2). Equation (4) checks if an event i has been completed on time by comparing the end timestamp TE_i with the planned end timestamp TP_i.

$$PPI_{DA,a} = \frac{\sum_i^E (x_{i,a} \times y_i)}{\sum_i^E x_{i,a}} \text{ with } PPI_{DA,a} \in [0,1] \; \forall a \tag{3}$$

$$y_i = \begin{cases} 1; \, TE_i - TP_i \leq 0 \\ 0; \, TE_i - TP_i > 0 \end{cases} \forall i \tag{4}$$

In step five, PPI for the trace and process perspectives are calculated. Equation (5) displays the calculation of $PPI_{PT,j}$ for the process time PT for the trace of an object j. Each object j has a trace with several events E_j. $PPI_{PT,j}$ is calculated based on two timestamps that are differentiated by two indices. The first index refers to the object of the trace, the second to the position of the event of this object in the trace. Consequently, $TS_{j,1}$ is the start timestamp of the first event in the trace of object j. TE_{j,E_j} is the end timestamp of the last event in the trace of object j.

$$PPI_{PT,j} = TE_{j,E_j} - TS_{j,1} \text{ with } PPI_{PT,j} \in [0,\infty) \; \forall j \tag{5}$$

Equation (6) is the calculation of the process time $PPI_{PT,p}$ for the process. The process has several traces T. The process time for the process perspective is the average of all process times of the traces in that process (see Eq. (5)).

$$PPI_{PT,p} = \frac{\sum_{j=1}^T PPI_{PT,j}}{T} \text{ with } PPI_{PT} \in [0,\infty) \tag{6}$$

Equation (7) shows the calculation for the deadline adherence $PPI_{DA,j}$ for the trace of an object j. The deadline adherence for a trace is the fraction of events in the trace, completed within the planned time frame. Equation (8) compares the end timestamp $TE_{i,j}$ with the planned end timestamp $TP_{i,j}$ of the event i within the trace of object j.

$$PPI_{DA,j} = \frac{\sum_{i=1}^{E_j} y_{j,i}}{E_j} \text{ with } PPI_{DA,j} \in [0,1] \; \forall j \tag{7}$$

$$y_{i,j} = \begin{cases} 1; \, TE_{i,j} - TP_{i,j} \leq 0 \\ 0; \, TE_{i,j} - TP_{i,j} > 0 \end{cases} \text{for } \forall i,j \tag{8}$$

Equation (9) presents the calculation for the deadline adherence $PPI_{DA,p}$ for the process. The deadline adherence for the process is the fraction of traces, of which the last event was completed within the planned timeframe. This is calculated using the expression in Eq. (8), whereby y_{j,E_j} compares the end timestamp TE_{i,E_j} with the planned end timestamp TP_{i,E_j} of the last event E_j within the trace of an object j.

$$\mathrm{PPI}_{\mathrm{DA},p} = \frac{\sum_{j=1}^{T} y_{j,E_j}}{T} \text{ with } \mathrm{PPI}_{\mathrm{DA},p} \in [0,1] \tag{9}$$

3.4 Detailed Description of Step Six

Step six of the methodology uses a discovery algorithm to map the process model of the ETEOPP. The aim is to create transparency of the ETEOPP and to put the calculated PPI into a context. As the popular discovery algorithms cannot deal with multiple OT, a DT of step two is chosen as input. Additionally, independent from the OT viewpoint of which the PP of the ETEOPP should be described and the PPI are calculated, and the input for step six must be the DT with the smallest sub-instance OT of the ETEOPP. In a manufacturing company, a product is represented by an article. The OT customer order might contain multiple articles per object, which disqualifies the DT of customer orders as input for step six. Otherwise, if products need to be manufactured one by one, the OT manufacturing order would contain exactly one article per object. The DT of manufacturing orders would then qualify to be selected as an input for step six. In industry, an OT, which contains one article per object, can be defined as the smallest sub-instance OT of the ETEOPP. The selection of the smallest sub-instance OT addresses the divergence problem on process discovery. The convergence problem persists, such that some process instances are duplicated when flattening towards the OT of an article, respective manufacturing order. As a result, the flattened DT has more events than the original process. This replication of events is acceptable because the PPI are calculated separately and the resulting process model does not display the number of events.

The herein used discovery algorithm is interchangeable as the selection of a suitable discovery algorithm depends on the requirements and data [21].

4 Introducing the Case Study and Validation of the Methodology

The methodology is validated with a dataset. The dataset is based on three order types (i.e. OT) processed through an exemplary ETEOPP depicted in Fig. 3. The process shows various tasks across the departments sales, manufacturing and shipping. The process includes parallel and sequential activities, OR-splits, AND-splits and loops of various lengths to test the robustness of the methodology.

The departments record their activities using different OT. The objects of all OT contains manufacturing orders as the smallest sub-instance as defined in Sect. 3.4. A manufacturing order only includes one article, but customer orders and shipping orders include one or more articles. Thus, customer orders and manufacturing

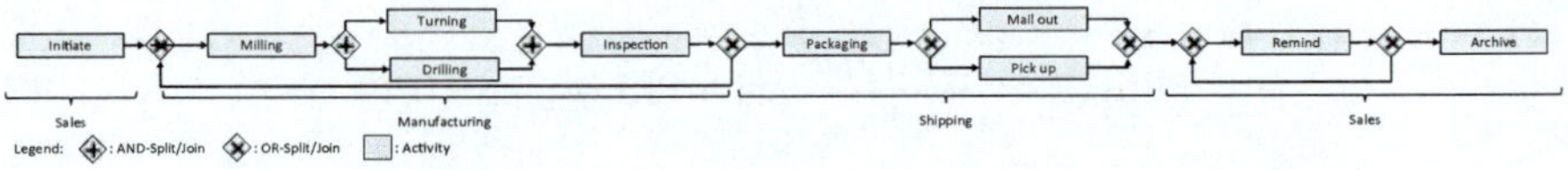

Fig. 3 BPMN model of the ETEOPP throughout three departments

orders are related one-to-many (1:*n*), which means that a customer order contains multiple manufacturing orders. Shipping order and manufacturing order are related many-to-one (*n*:1), which means that multiple manufacturing orders are shipped in the same shipping order. The OT customer order and shipping order are related many-to-many (*n*:*n*). In practice, two customer orders are shipped to the same address across three shipping orders.

The OT is processed in the ETEOPP, as shown in the conceptual object-centric Petri net in Fig. 4. An object-centric Petri net extends a regular Petri net by shading transitions and places based on the OTs they refer to. Places and arcs of transitions consuming multiple objects are highlighted by double lines [14]. Due to the incomprehensibility of object-centric Petri nets in practice, more intuitive visualizations and established process models (respective miners) are used for the case study.

The dataset comprises three DTs with 41 events involving two customer orders, five manufacturing orders and three shipping orders. Table 1 shows the first line of the DT from the manufacturing department. Here, the customer order is recorded as data for every activity.

Tables 2, 3 and 4 show the resulting PPI for the process time and deadline adherence for selected activities, objects and OTs based on the Eqs. 1–9. The process time of traces is large compared to the process time of activities, partly because time outside of work shifts were not excluded from the calculations.

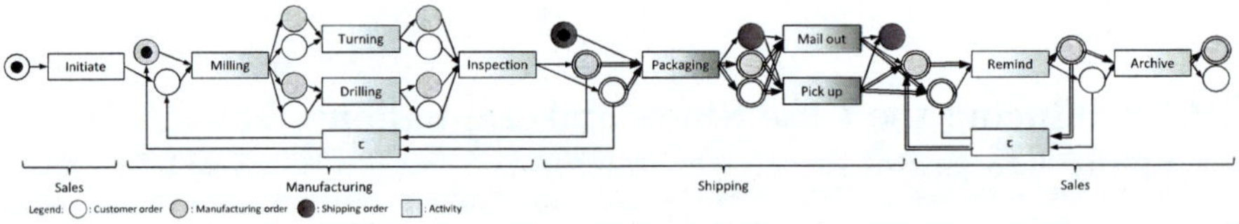

Fig. 4 Object-centric Petri net of the ETEOPP, showing the OTs treated in each activity

Table 1 First lines of the DT from the manufacturing department

Manuf. order	Activity	Start	End	Planned end	Order received	Customer order
M378	Milling	11.02.20 09:00	11.02.20 14:40	11.02.20 15:00	05.02.20 14:10	13,623
…	…	…	…	…	…	…

Table 2 PPI for the Milling activity

Activity	Process time [h]	Deadline adherence
Milling	6.73	0.50

Table 3 PPI for the trace of M28910 for the OT manufacturing order

Object of the traces	Process time [h]	Deadline adherence
M28910	773.33	0.54

Table 4 PPI of the process for the OT manufacturing order and customer order

Process	Process time [h]	Deadline adherence
Customer order	738.67	0.50
Manufacturing order	731.73	0.60

Figure 5 shows the process model, which was mapped using the DT of the OT manufacturing order. This DT was chosen according to Sect. 3.4, and the resulting process model is valid for evaluating PP independent of the OT chosen to calculate PPI. The event log was extended for process discovery to include 123 events to approximate a bigger dataset. The software ProM 6.9 and the plug-in *Mine process tree with Inductive Miner* followed by the plug-in *Convert Process tree to BPMN diagram* were used to describe the ETEOPP process model. The process model is under-fitting. The activity *inspection* is a successor of the activity *initiate*, which is not possible in the real process. One reason is the inductive miner and its trade-off between under-fitting process models and preserving fitness. Here, the heuristic miner was able to produce a better fitting process model (see Fig. 6). For this, the plug-ins *Heuristic net*, *Convert Heuristic net into Petri net* and *Convert Petri net to a BPMN diagram* were applied.

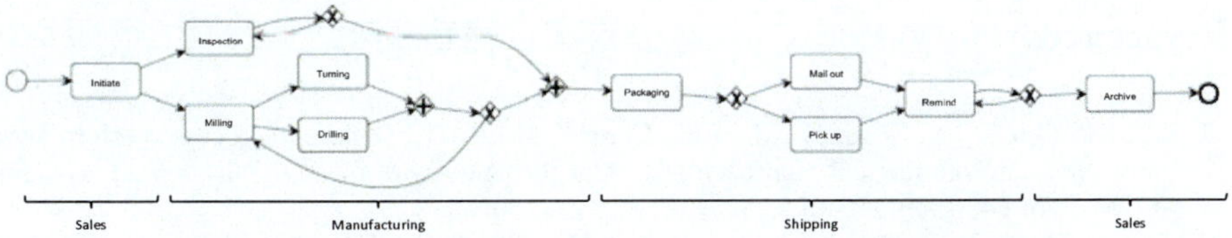

Fig. 5 Process model mapped using the inductive miner and the OT manufacturing order

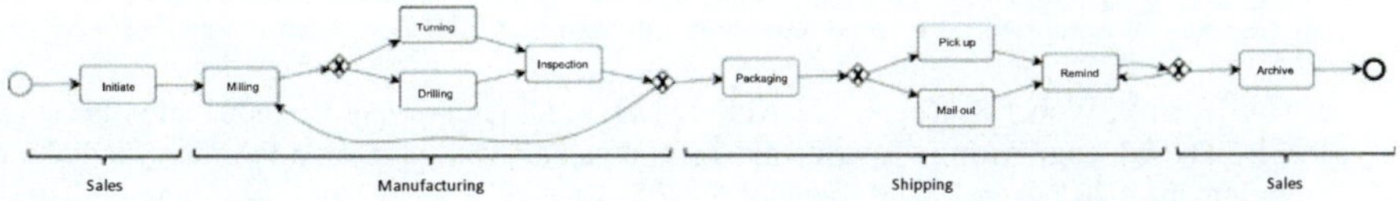

Fig. 6 Process model mapped using the heuristic miner and the OT manufacturing order

5 Summary and Research Outlook

This paper demonstrated a methodology for the application of PM in ETEOPP. The six contributing steps merge event logs from companies' information systems into an E2EL and use the results for calculating PPI and discovering the process model. The novelties are the consolidation of multiple event logs of the ETEOPP and the use of an OCL to deal with multiple order types in production companies in the context of PM. Thus, analysis of the ETEOPP can be based on facts and exempt from employees' subjectivity and other external factors. This enables long-term and continuous improvement of PP in projects commencing with the description of as-is PP. An application of the methodology on a dataset results in a visualisation of the ETEOPP process model and calculated PPI.

The presented methodology expands the second step of a broader approach presented in [3]. As an outlook, the preceding and subsequent steps of the broader approach need to be elaborated before integrating the separate parts into a holistic solution for describing PP in ETEOPP. In particular, an approach for defining the data requirements for the DT from software infrastructure and a user interface to operate the methodology and display the results needs to be developed. Furthermore, applications with real company data would have the potential of uncovering development potential.

Next, a data-based approach for processing acquisition can always be assisted by classical participative methods since it helps detecting hidden activities or inefficiencies and further improvement potentials that are not stored in a company's software infrastructure.

Acknowledgements Funded by the Deutsche Forschungsgemeinschaft (DFG, German Research Foundation) under Germany's Excellence Strategy—EXC-2023 Internet of Production—390621612.

References

1. Del-Río-Ortega A, Resinas M, Ruiz-Cortés A (2010) Defining process performance indicators: an ontological approach. In: On the move to meaningful internet systems (OTM), vol 6426, pp 555–572
2. Dumas M, La Rosa M, Mendling J, Reijers HA (2018) Fundamentals of Business Process Management. Springer, Berlin Heidelberg
3. Schuh G, Gützlaff A, Schmitz S, van der Aalst W (2020) Data-based description of process performance in end-to-end order processing. CIRP Ann Manuf Technol 69:381–384
4. Eversheim W, Krumm S, Heuser T, Müller S (1993) Process-oriented organization of order processing—a new method to meet customers demands. CIRP Ann Manuf Technol 42:569–571
5. Schönsleben P, Weber S, Koenigs S, Aldo D (2017) Different types of cooperation between the R&D and engineering departments in companies with a design-to-order production environment. CIRP Ann Manuf Technol 66:405–408

6. Schuh G, Gützlaff A, Cremer S, Schopen M (2020) Understanding process mining for data-driven optimization of order processing. In: Conference on Learn Factories, pp 417–422
7. Harmon P, Garcia J (2020) The BPTrends report. In: The state of business process management: 2020
8. Reinkemeyer L (2020) Process mining in action. Springer, Berlin Heidelberg
9. Pospísil M, Mates V, Hruska T, Bartik V (2013) Process mining in a manufacturing company for predictions and planning. Int J Adv Softw 6(3 & 4):283–297
10. Van der Aalst W (2016) Process mining. Springer, Berlin Heidelberg
11. Van der Aalst W, Alves de Medeiros AK, Weijters AJMM (2005) Generic process mining. In: Proceedings of 26th international conference on applications and theory of petri nets, pp 48–69
12. Thaler T, Ternis S, Fettke P, Loos P (2015) A comparative analysis of process instance cluster techniques. In: AISeL Wirtschaftsinformatik Proceedings, pp 423–437
13. Günther CW, Verbeek E (2014) XES standard definition, TU Eindhoven
14. Van der Aalst W, Berti A (2019) Discovering object-centric petri nets. In: Fundamenta Informaticae XXI, pp 1001–1042
15. Marr B (2015) Big data: using SMART big data, analytics and metrics to make better decisions and improve performance. John Wiley & Sons, Hoboken NJ USA
16. Van der Aalst W et al (2012) Process mining manifesto. In: Business process management workshops. BPM 2011. Lecture Notes in Business Information Processing, pp 169–194
17. Van der Aalst W (2019) Object-centric process mining: dealing with divergence and convergence in event data software engineering and formal methods, SEFM, pp 3–25
18. Leemans S, Poppe E, Wynn MT (2019) Directly follows-based process mining: exploration & a case study. In: International conference on process mining (ICPM), pp 25–32
19. Zaki NM, Awad A, Ezat E (2015) Extracting accurate performance indicators from execution logs using process models. In: Proceedings of IEEE/ACS 12th International Conference of Computer Systems and Applications, vol 1, pp 1–8
20. Schmitz S, Renneberg F, Cremer S, Gützlaff A, Schuh G (2020) Definition of process performance indicators for the application of process mining in end-to-end order processing processes. In: Proceedings of 10th congress of the German academic association for production technology, pp 670–679
21. Jouck T, Bolt A, Depaire B, Massimiliano de Leoni, van der Aalst W (2018) An integrated framework for process discovery algorithm evaluation. In: IEEE transactions on knowledge and data engineering. arXiv:1806.07222v1

Performance Testing and Preventive Maintenance of Automatic Handwashing Tool in the Efforts of COVID-19 Prevention

Adhan Efendi, Rosiah, and Ade Nuraeni

1 Introduction

The coronavirus pandemic that spread in early 2020 is a nightmare for all countries in the world. This virus is overwhelming for all countries to prevent more people from becoming infected. It is due to the rapid spread of the virus from person to person. The city of Wuhan, Hubei Province in China became the first place where the pneumonia virus was discovered, which is indicated to have originated from one of the animals sold in the city's market [1]. The speed of spread and the alarming rate of death have led many countries and jurisdictions to introduce measures to prevent the spread of COVID-19 [2]. Reference [3] added that the spread of the virus is also influenced by a person's antibodies, if the antibodies are good, the risk of virus attack is also getting smaller.

Washing hands is one of the potent steps in preventing disease during a pandemic. Other factors, namely the availability of clean water, sanitation, and hygienic conditions are absolute to protect human health during all outbreaks of infectious diseases, including the 2019 coronavirus disease [4]. The patent on handwashing tool was developed by [5], a handwashing device made with a simple sensor system, so that water spraying when washing hands is done automatically. Research by [6] developed a handwashing device with a pedal system that is stepped on in the handwashing process. Washing hands is one of the factors to prevent the spread of the coronavirus, but it is also necessary to use masks as an individual effort to maintain their health [7]. In addition to handwashing tools, people also need to know the correct handwashing technique [8]. Added by [9] washing hands is a simple activity, but often many individuals are negligent in implementing these healthy behaviors.

Subang State Polytechnic is one of the vocational education campuses in West Java, Indonesia [10]. This research is a continuation of previous research that

A. Efendi (✉) · Rosiah · A. Nuraeni
Subang State Polytechnic, Subang 41211, Indonesia

R. K. Agarwal (ed.), *Recent Advances in Manufacturing Engineering and Processes*, Lecture Notes in Mechanical Engineering,
https://doi.org/10.1007/978-981-16-3934-0_16

focuses on discussing the design and construction of automatic handwashing devices. Furthermore, this research will focus on discussing the performance test of automatic handwashing tool that will be implemented at one of the public health centers in Subang Regency. Continuous use of the tool will cause the tool to quickly experience damage. So, it is necessary to carry out preventive maintenance to maintain the performance of handwashing tools. Added by [11] required preventive type of maintenance for newly used machines in an effort to maintain engine condition in the best performance. Preventive maintenance is needed by large-scale industry or small-scale industry [12].

2 Method

This study is a kind of research and development. Data is collected by observation and documentation, then the data is analyzed descriptively qualitatively. The stages of this research include reading the design drawings and manufacturing results, then conducting trials of automatic handwashing tools, observation and identification of components, making preventive maintenance schedules, expert validation, and implementing preventive lists.

3 Result and Discussion

3.1 Design and Manufacture

The automatic handwashing tool begins with a tool design carried out by the research team using the inventor Autodesk application. After the design complete, the researcher carries out the manufacturing process. Execution images through the process of observation, sketching tools, making 2-dimensional and 3-dimensional images. Researchers also designed an analysis of the component position and the load received when entered the manufacturing process. Design automatic handwashing can be seen in Fig. 1.

The manufacturing process includes measuring, cutting, joining, and painting. An automatic handwashing tool uses the sun's heat as its energy source. This tool uses solar cells so that solar energy is absorbed and stored in the battery. This handwashing tool uses osmosis as a filter to take advantage of recycled water used in the first handwashing process. This tool also uses a motion sensor on the water faucet to reduce touch on the tap and maintain distance between users more effectively. The final step is to paint and test the performance of the automatic handwashing tool. It aimed to determine the battery power capacity, the running of the pump, and the function of the sensors installed on the device.

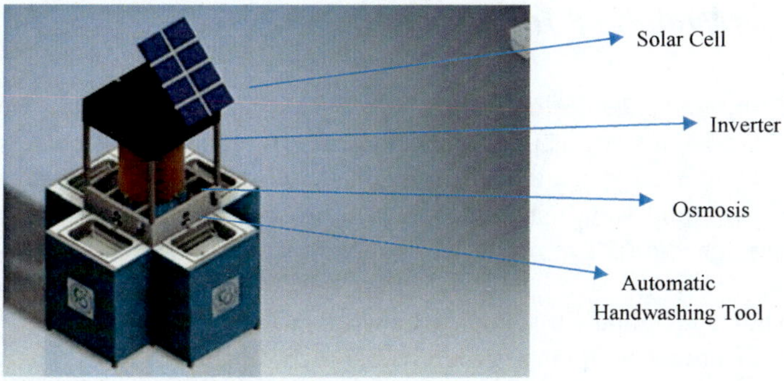

Fig. 1 Design automatic handwashing tool

3.2 *Performance Testing*

Testing is intended to see the performance of the handwashing tools made. Testing aims to look at the performance of each component incorporated in the unit. The test was carried out at the Kasomalang Subang Community Health Center. The results obtained in the first stage of testing, namely handwashing tools can be used evenly for 6 h. After 6 hours of use, the water in the gallon must be replenished. This is because the water from the osmosis is so small that the filling of water into gallon is very slow. The results of the second test resulted in the need for scheduling of recycled water handwashing equipment components. This is because, during a pandemic, handwashing equipment is always used. So, it is necessary to make a routine tool maintenance schedule to maintain the condition of the equipment in optimal condition. The next test results show that handwashing tools are optimal if the device terminates when not in use. It can protect the battery so that the tools can operate for more than 6 h once. Performance testing automatic handwashing can be seen in Fig. 2.

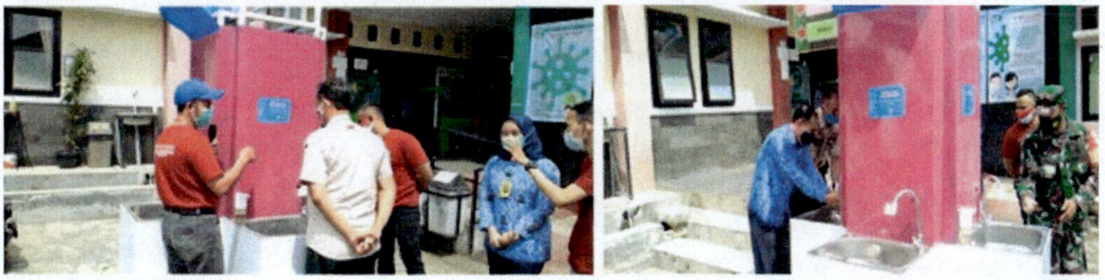

Fig. 2 Performance testing

3.3 Observation and Identification

Making a routine schedule for automatic recycled handwashing tool begins with observing and identifying handwashing tools. Observation is intended to classify components that need to be maintained regularly. The identification is needed in the process of preparing what material tools will be used in implementing preventive maintenance schedules. The results of observation and identification are displayed in Table 1.

The Table 1 concludes that four groups of main components must be treated, such as the electrical component group which according to the risk is categorized as red, meaning that this group has a high risk of damage. Recycled component groups such as osmosis and water collection boxes are in the yellow risk category, meaning that the component group that has a high enough risk of being damaged. Finally, there is a group of power and case components that are in the green risk category, meaning that these components are sufficiently inspected in terms of performance monitoring that might cause interference.

3.4 Schedule Maintenance and Expert Validation

Preventive maintenance is routine maintenance that aims to keep the tool in optimal condition. Making a schedule for recycled automatic handwashing tool using the inspection, small repair, medium repair, and overhaul (ISMO) method. The schedule plan was then validated by an expert. The maintenance schedule that has been approved by the expert can be seen in Table 2.

Table 1 Table observation and identification

Tool parts	Maintenance schedule plan	Risk information
Electricity system (battery, inverter, sensor, and cable)	Weekly maintenance (inspection, small repair, and medium repair)	
Recycling system (osmosis, water collection box)	Weekly maintenance (inspection and small repair)	
Power system (solar cell)	Monthly maintenance (inspection)	
Hand wash case	Monthly maintenance (inspection)	

Table 2 Schedule maintenance

No.	Tool parts	Monthly maintenance				Weekly maintenance			
		I	S	M	O	I	S	M	O
1	Battery					5’			
2	Inverter					5’	10’	15’	
3	Sensor					5’	10’		
4	Cable					5’			
5	Osmosis	10’	25’						
6	Water collection box	10’	25’						
7	Solar cell	15’							
8	Hand wash case	15’							
Total		50’	50’			20’	20’	15’	
						155’			

Based on Table 2, it is explained that the preventive maintenance schedule that has gone through expert validation uses the ISMO method with a monthly and weekly routine schedule. The total time required is 155 min. Note from the validator that the level of inspection is continued to the overhaul level if, in the medium repair stage, the machine condition is still experiencing significant problems.

4 Result and Discussion

Based on the data and research that has been done, it can be concluded that the results showed the performance test of the handwashing tool to be able to use within 6 h of use. The next test results show that handwashing tools are optimal if the device terminates when not in use. It can protect the battery so that the tools can operate for more than 6 h once. It is very important to develop a high storage capacity of battery so the device can work more optimal. Preventive maintenance with monthly and weekly routine schedules using inspection, small repair, medium repair, and overhaul methods. The total time needed for precaution maintenance of handwashing equipment is 155 min. For future research, it is also advisable to develop a corrective schedule for automatic handwashing. A combination of preventive and corrective maintenance is highly recommended for tools that are continuously used on a regular basis. The implementation of a routine maintenance schedule for automatic handwashing tool that has been carried out gives an idea to us that the use of automatic handwashing tools is still used today as a measure to prevent the COVID-19 virus. With the availability of automatic handwashing tool, it is expected that it forms a habit in the community regarding the necessity of maintaining personal hygiene and preventing the spread of infectious diseases.

References

1. Muhammad K, Ahmed A, Rahman H, Khan Z (2020) Coronavirus outbreaks: prevention and management recommendations. Drugs Therapy Perspect 1(March 2020):1–4
2. Alzyood M et al (2020) COVID-19 reinforces the importance of hand washing. J Clin Nurs 29 (April 2020):15–16
3. Chen RC, Zhong NS, W-J Guan (2020) Strategies for the prevention and management of coronavirus disease 2019. Euro Respir J 55(4):1563–1569
4. World Health Organization (2020) Water, sanitation, hygiene, and waste management for the COVID-19 virus. Unicef, United State
5. Newell RE (1962) Hand washing machine. United States Patent 3081471, 16 07 1962
6. Tiruneh AT (2016) Development of "press tap" hand washing technology: design, construction and operational aspects. East Asian Sci Technol Soc Int J 4(3):48–56
7. Peters A, Pittet D, Lotfinejad N (2020) Hand hygiene and the novel coronavirus pandemic: the role of healthcare workers. J Hos Infect 105(4):1–5
8. Olorunnishola AA, Osayomi PA, Ejiko SO (2020) Ergonomic development of manual hand washing machine to ameliorate the deadly effect of COVID 19 pandemic. Int Adv Res J Sci Eng Technol 7(7):6–16
9. Ray I (2020) Viewpoint—handwashing and COVID-19: simple, right there…? World Dev 135(2020):105086–105088
10. Nugraha A, Baharta R, Efendi A (2019) Manufacturing of electrical dryer machine for food and fruit products. In: IOP conference series: materials science and engineering, China
11. C. U.S. Department of the Interior Bureau of Reclamation Denver. Maintenance Scheduling for Mechanical Equipment. U.S. Department of the Interior Bureau of Reclamation Denver, Colorado, United States, 2009
12. Yogish H et al (2016) Preventive maintenance and breakdown reduction of critical machines. Int Res J Eng Technol (IRJET) (United States)

On the Unimportance of the Number and the Type of Elements While Solving Certain Problems

P. Kirana Kumara and Kota Ramakanth Yasaswi

1 Introduction

Real-time simulation of biological organs may require fast finite element simulations [1, 2]. It is a known fact that if we increase the total number of elements or/and use a better element, the results would be more accurate [3, 4], but the computational time would also increase. Although it is good to get more accurate solutions, increased solution time is not desirable [5]. Hence it is better if one could achieve faster simulations without practically sacrificing accuracy, and this is possible if the simulation results do not depend (practically) on the type of the element used and/or the number of elements used.

Biological organs show non-linear material behavior. However, many of the real-time simulations of biological organs assume that they are linear elastic (small displacement/strain also), to be able to achieve fast simulations. Hence this work tries to find out whether it is true that the solutions are practically independent of the type and/or the number of elements used, at least for some problems, assuming linear-elastic material behavior (including small strain/displacement).

The importance of real-time simulations is very evident in the literature [6, 7]. The reference [8] notes that the results obtained are almost the same irrespective of whether geometric nonlinearity (with linear-elastic constitutive model) is assumed or whether the material is assumed to follow hyperelastic behavior, while computing brain deformation for image-guided surgery. However, one cannot find any source in the literature that solves a set of problems using different element types and different mesh resolutions (different number of elements) to demonstrate that the element types and the total number of elements used do not significantly affect the accuracy of the results, at least for the problems taken up for study.

P. Kirana Kumara (✉) · K. R. Yasaswi
Department of Mechanical and Manufacturing Engineering, Manipal Institute of Technology, Manipal Academy of Higher Education, Manipal, Karnataka 576104, India

R. K. Agarwal (ed.), *Recent Advances in Manufacturing Engineering and Processes*, Lecture Notes in Mechanical Engineering,
https://doi.org/10.1007/978-981-16-3934-0_17

2 Methodology

In this work, material is always assumed linear elastic (small displacement/strain also). The commercial software package ANSYS is used for solving the problems.

As for the structural equations used (or the physics behind the problems taken up in this work), the equations for small displacement/strain linear elasticity are well-known and are incorporated in ANSYS. Since this author is not developing his own code for this work, the equations (for small displacement/strain linear elasticity) are not mentioned here. The elements used in this work are "Tet 4 node 285" and "Tet 10 node 187" (in ANSYS), and both support small displacement/strain linear elasticity.

Simulations are carried out for three problems, named here as Problem 1, Problem 2, and Problem 3.

Problem 1 and Problem 2 both consider a cantilever beam of 100-mm length and 1 mm by 1 mm cross section. Young's modulus of the material is assumed to be equal to 210 GPa, for both the problems. For both the problems, the left end of the beam is completely fixed. For each of Problem 1 and Problem 2, problem is solved using three mesh resolutions (using a total of 10 elements, 30 elements, and 500 elements). For each of the mesh resolutions, two types of elements are used. The elements used are: 2 noded beam element (3D finite strain 2 node 188, in ANSYS) and 3 noded beam element (3D finite strain 3 node 189, in ANSYS). Hence for each of Problem 1 and Problem 2, there are six simulations in total (three mesh resolutions and two types of elements). Deformations of the cantilever beam for each of the six simulations (separately for Problem 1 and Problem 2) are plotted as graphs (these graphs are generated by ANSYS). Finally, for the sake of comparison, all the six deformed shapes are plotted on just one graph, separately for each of Problem 1 and Problem 2.

For Problem 1, the right end is subjected to a displacement of 20 mm in the downward direction. For Problem 2, the right end is subjected to a force of 1.04958 N in the downward direction. The reason for applying a force of 1.04958 N at the right end of the beam (for Problem 2) is that if one uses the analytical solution given by the simple beam theory (pure bending) [9, 10], a force of 1.04958 N at the right end of the beam would result in a displacement of 20 mm at the point of application of load (the same amount of displacement at the right end, as that for Problem 1).

Problem 1 and Problem 2 do not have any biomechanical significance. They are used here just to demonstrate that the results could be practically independent of the type and number of elements used (at least for some problems), even for different types of boundary conditions (applied displacement or applied force). Plotting the deformation of cantilever beam is a popular benchmark problem.

Problem 3 is to consider a sphere of 150-mm diameter. The reason for choosing 150-mm diameter is that this value is of the same order as that of the dimensions of some biological organs such as human liver or kidney [11–13]. Young's modulus of the material is assumed to be equal to 150 MPa, and the Poisson's ratio is assumed to be equal to 0.4. These values of the material constants are of the same order as that of some biological organs such as human liver or kidney. A fine mesh is used to discretize the domain. Three nodes that are on the surface of the sphere and that are adjacent to one another are completely fixed (all degrees of freedom arrested), and a displacement of 30 mm is applied in all of x, y, and z directions, on a node that is located on the other side of the sphere (compared to the three nodes that are completely fixed). The problem is solved using two types of solid elements: 4 noded tetrahedral element (Tet 4 node 285, in ANSYS), and 10 noded tetrahedral element (Tet 10 node 187, in ANSYS), and the undeformed and deformed shapes are separately plotted for both the simulations.

3 Results

The deformed shapes for the six simulations corresponding to Problem 1 are shown in Figs. 1, 2, 3, 4, 5 and 6. The titles of the figures convey the type and the number of elements used.

The deformed shapes for the six simulations corresponding to Problem 2 are shown in Figs. 7, 8, 9, 10, 11 and 12. The titles of the figures convey the type and the number of elements used.

Figure 13 plots the deformed shapes corresponding to all the six simulations (corresponding to Problem 1) on a single graph. One may note that the deformed shapes are almost identical and hence although there are six deformed shapes on the same graph, it looks like just one deformed shape is plotted; this clearly shows how well the deformed shapes match; the deformed shapes are almost the same irrespective of the type or the number of elements used for solving Problem 1.

Figure 14 plots the deformed shapes corresponding to all the six simulations (corresponding to Problem 2) on a single graph. One may note that the deformed shapes are almost identical and hence although there are six deformed shapes on the same graph, it looks like just one deformed shape is plotted; this clearly shows how well the deformed shapes match; the deformed shapes are almost the same irrespective of the type or the number of elements used for solving Problem 2.

The deformed shape corresponding to Problem 3, solved using the 4 noded tetrahedral elements is shown in Fig. 15. Similarly, the deformed shape corresponding to Problem 3, solved using the 10 noded tetrahedral elements is shown in Fig. 16.

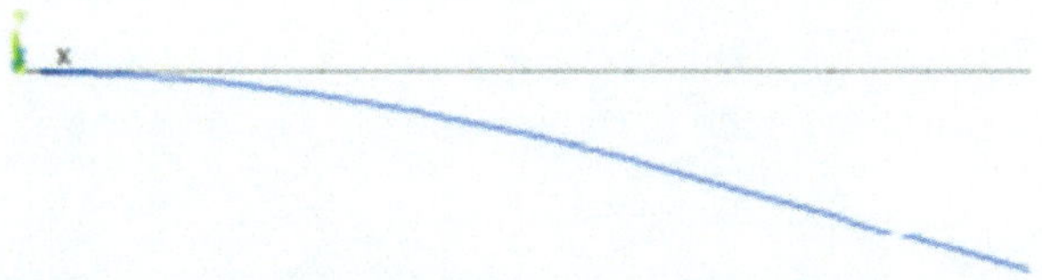

Fig. 1 Problem 1, 2 noded beam, 10 elements

Fig. 2 Problem 1, 3 noded beam, 10 elements

Fig. 3 Problem 1, 2 noded beam, 30 elements

Fig. 4 Problem 1, 3 noded beam, 30 elements

Fig. 5 Problem 1, 2 noded beam, 500 elements

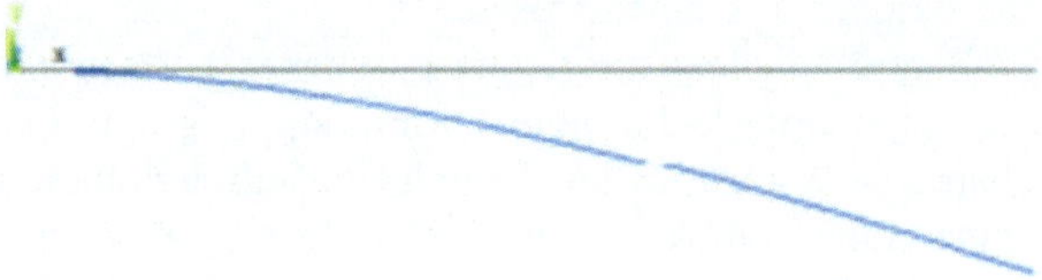

Fig. 6 Problem 1, 3 noded beam, 500 elements

Fig. 7 Problem 2, 2 noded beam, 10 elements

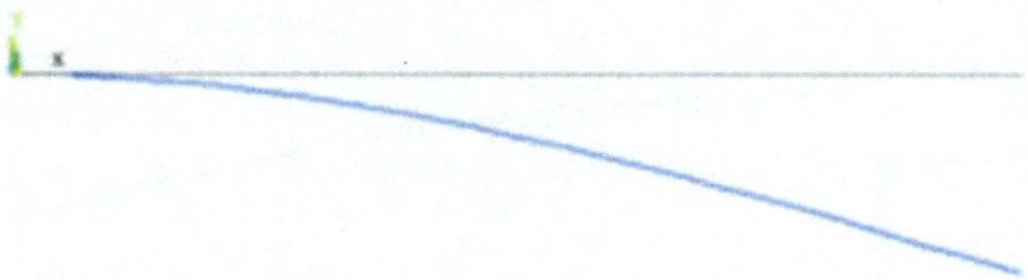

Fig. 8 Problem 2, 3 noded beam, 10 elements

Fig. 9 Problem 2, 2 noded beam, 30 elements

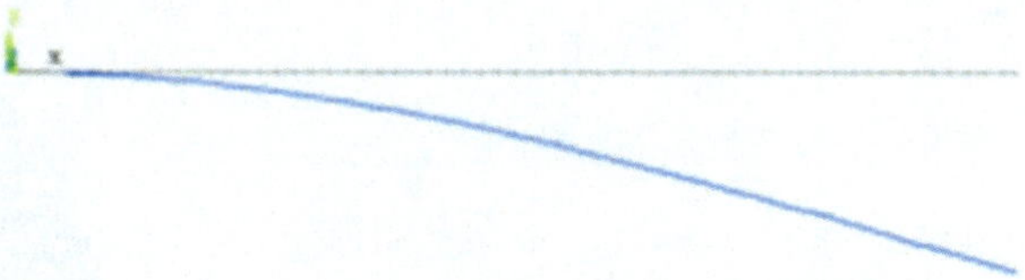

Fig. 10 Problem 2, 3 noded beam, 30 elements

Fig. 11 Problem 2, 2 noded beam, 500 elements

Fig. 12 Problem 2, 3 noded beam, 500 elements

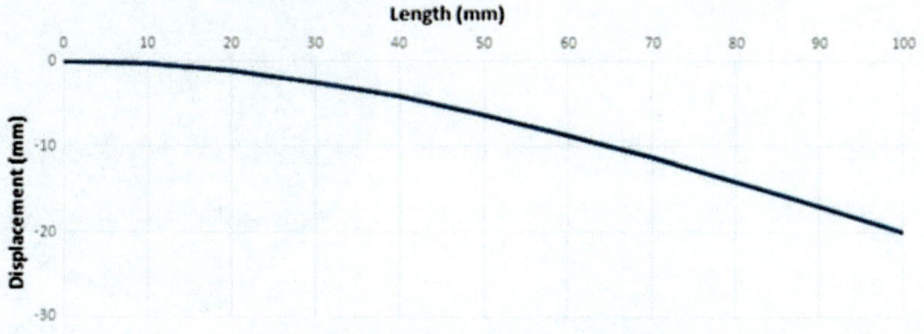

Fig. 13 All deformed shapes for Problem 1

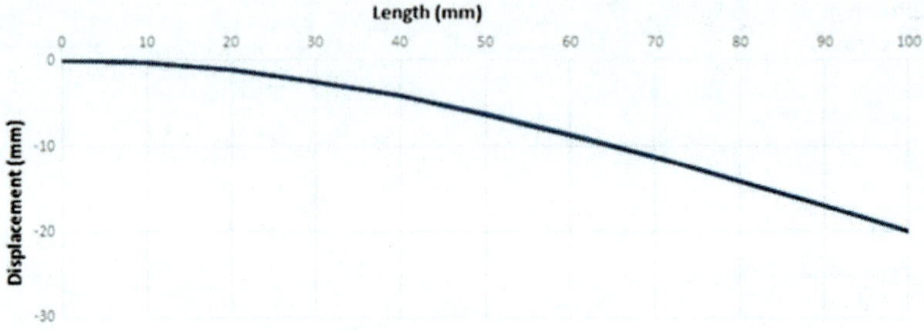

Fig. 14 All deformed shapes for Problem 2

Fig. 15 Problem 3, deformed shape, 4 noded tetrahedral element

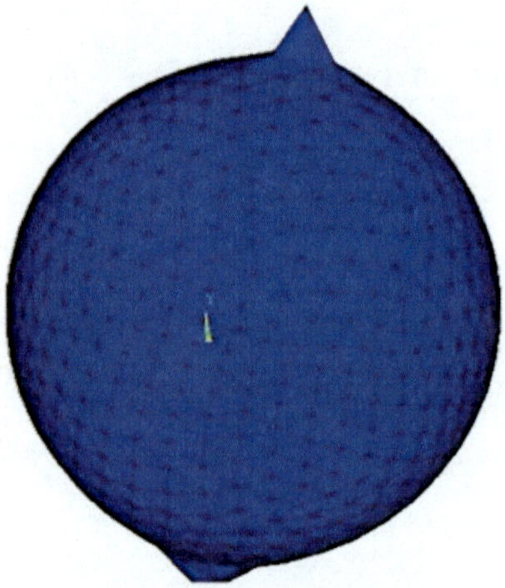

Fig. 16 Problem 3, deformed shape, 10 noded tetrahedral element

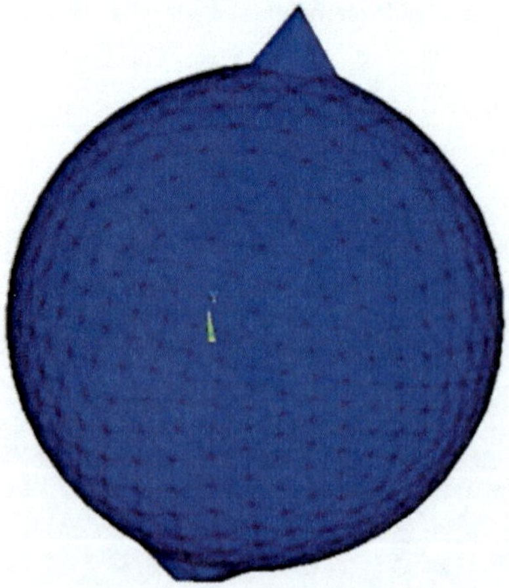

4 Discussion

From graphs, one can see that all the six deformed shapes plotted for Problem 1 look almost the same. This is even clearer from graph where all the six deformed shapes are plotted on the same graph. Since the six deformed shapes correspond to different element types and mesh resolutions, and since all the six plots look almost the same, one may infer that the results obtained are practically the same irrespective of the type or number of elements used, for Problem 1.

Similarly, all the six deformed shapes plotted for Problem 2 also look almost the same. This is also clear from graph where all the six deformed shapes are plotted on the same graph. For Problem 2 also, the six deformed shapes correspond to different element types and mesh resolutions, and since all the six plots look almost the same, one may infer that the results obtained are practically the same irrespective of the type or number of elements used.

Finally, one can see that the deformed shapes are almost the same irrespective of the type of elements used for the simulations corresponding to Problem 3 also.

5 Conclusion

For the problems considered in this work, the results (deformed shapes) do not noticeably depend on the type or the number of elements used to solve the problems.

One may note that the present work is novel from two perspectives: (i) it recognizes that results could be practically independent of the number or type of elements used, for certain problems (ii) it recognizes that real-time simulation of biological organs could benefit from the fact that results could be practically independent of the number or type of elements used, for certain problems.

As future work, more number of problems (of new types) could be solved using different types and numbers of elements to confirm whether or not the results can significantly vary for different types and numbers of elements for these new problems.

References

1. Meier U, Lopez O, Monserrat C, Juan MC, Alcaniz M (2005) Real-time deformable models for surgery simulation: a survey. Comput Methods Programs Biomed 77:183–197. https://doi.org/10.1016/j.cmpb.2004.11.002
2. Delingette H (1998) Toward realistic soft-tissue modeling in medical simulation. Proc IEEE 86(3):512–523

3. Kirana Kumara P (2020) Simulation of mechanical behaviour of materials using constant boundary elements—a discussion on the accuracy of results for bars. Mater Sci Forum (ISSN: 1662–9752) 1013:99–105. https://doi.org/10.4028/www.scientific.net/MSF.1013.99
4. Kirana Kumara P (2020) Simulation of mechanical behaviour of materials using constant boundary elements—a discussion on the accuracy of results for beams. Mater Sci Forum (ISSN: 1662–9752) 1013:106–113. https://doi.org/10.4028/www.scientific.net/MSF.1013.106
5. Kirana Kumara P (2016) Studies on the viability of the boundary element method for the real-time simulation of biological organs. PhD Thesis, Centre for Product Design and Manufacturing, Indian Institute of Science, Bangalore, India
6. Kirana Kumara P (2014) A study of speed of the boundary element method as applied to the realtime computational simulation of biological organs. Electron J Boundary Elem 12(2):1–25
7. Kirana Kumara P, Ghosal A (2012) Real-time computer simulation of three dimensional elastostatics using the finite point method. Appl Mech Mater 110(16):2740–2745
8. Wittek A, Hawkins T, Miller K (2009) On the unimportance of constitutive models in computing brain deformation for image-guided surgery. Biomech Model Mechanobiol 8:77–84
9. Young WC, Budynas RG (2002) Roark's formulas for stress and strain, 7th edn. McGraw-Hill
10. Mahadevan K, Balaveera Reddy K (2000) Design data hand book, 3rd edn. CBS Publishers and Distributors, New Delhi
11. Gray H (1918) Anatomy of the human body. Lea & Febiger, Philadelphia
12. Kirana Kumara P (2012) Extracting three dimensional surface model of human kidney from the visible human data set using free software. Leonardo Electron J Practices Technol 11 (20):115–126
13. Kirana Kumara P, Ghosal A (2010) A Procedure for the 3D reconstruction of biological organs from 2d image sequences. In: International conference on biomedical engineering and assistive technologies (BEATs-2010), 17–19 December 2010. Dr. B R Ambedkar National Institute of Technology, Jalandhar, India

Materials Science and Chemical Engineering

Evaluation the Effect of Emulsion on Pour-Point Reduction of Oil from Fang Oilfield

Tanapol Ruengnam and Kreangkrai Maneeintr

1 Introduction

The flow assurance is the important aspect for preventing economic loss in the petroleum industry. It contributes a lot in energy sector which its demand increases every years. Waxy crude oil is the one of major issues of flow assurance. Paraffin in crude oil is the most contribution for wax deposition [1]. The temperature that wax begins to participate out of the crude oil is wax appearance temperature (WAT). On the other hand, pour point is defined as the lowest temperature at which the fluid cannot be poured under that conditions. Therefore, the flow behaviour at the temperature between WAT and pour point temperature can lead to the wax deposition [2]. The two points are important for determination of crude oil properties [3]. For wax mitigation and remediation, there are three methods that are commonly used. The first one is thermal; then, mechanical methods and the last one is chemical methods [4]. The chemical method is the choice for this study because, unlike other techniques, it does not need to stop operation [5]. In the production which crude oil goes up to the surface, the waxy crude oil participates out of fluid and deposits. Therefore, the solvent is applying the chemical usage to reduce pour point to maintain fluid flowability. The selection of chemical is depended on the ability to dissolve, environmental impact, economic feasibility and availability. Another aspect is flammability of solvent which is mostly defined as flash point to prevent the combustibility of solvent [5]. In this research, there are four chemicals to study such as n-heptane being one of the best choices for viscosity reduction and increasing solubility of paraffin [6]. The second chemical is n-hexane which has the same attribute with the first one. Additionally, these two chemicals are solvent inhibitor for

T. Ruengnam · K. Maneeintr (✉)
Carbon Capture, Storage and Utilization Research Group, Department of Mining and Petroleum Engineering, Faculty of Engineering, Chulalongkorn University, Bangkok 10330, Thailand
e-mail: Krengkrai.M@chula.ac.th

R. K. Agarwal (ed.), *Recent Advances in Manufacturing Engineering and Processes*, Lecture Notes in Mechanical Engineering,
https://doi.org/10.1007/978-981-16-3934-0_18

wax prevention. Next, the third and fourth are crystal modifier comprising of Poly (maleic anhydride-alt-1-octadecene) which also called PMAO which can interfere the growth of crystal [7] and ethanolamine called MEA. In oil production, it is normally recovered crude oil with produced water. Furthermore, when there are two elements consisting of emulsifying agent which lower interfacial tension and degree of mixing. Emulsion is formed as two immiscible liquid mix together because of emulsifier. Therefore, dispersed phase of one liquid in continuous phase of another liquid is formed [8]. Also, emulsion can be found at every stages in oil recovery and process. There are 3 types of emulsions, water-in-oil, oil-in-water and water-in oil-in water or oil-in-water-in-oil. In the oilfield, it is normally found water-in-oil emulsions and the highest possible percentage of water volume fraction is high as 60%. Emulsion can cause the problem of fluid flow with high viscosities; thus, causing the flow assurance aspect [9]. Therefore, the objectives of this study are to select potential chemicals for pour point reduction and to investigate the effect of chemicals in crude oil and its emulsion on pour point reduction. From this study, the problem of wax deposition from crude oil production can be lowered with the pour point reducing agent such as n-heptane and this can be applied in the field for practical oil production in the future.

2 Materials and Methods

2.1 *Materials and Equipment*

Crude oil sample is obtained from Fang oilfield, Northern Thailand. The original pour point of crude oil is 36.3 °C. The chemicals, n-heptane, n-hexane, PMAO and MEA are purchased from Sigma-Aldrich. The experiment are conducted in the water bath with glycol as a heating media. The glass tubing with cock connected with the probe of thermocouple and experimental stand is used in this experiment.

2.2 *Pour Point Measurement*

Pour point measurement follows the standard procedure of ASTM D5853-11. It is conducted by varying water content with distilled water at 0, 20, 40 and 60% by volume mixing with crude oil to form emulsion. An electric mixer is used to mix for 15 min at room temperature. Each pour point measurement uses 15 cm^3 of emulsion ranging various condition and water content varying from 0 to 20% by volume for the chemical selection for pour point reduction. Also, the solution concentration is from 5, 10, 15 and 20% for hexane and heptane. Moreover, the concentration of the crystal modifiers is 2500, 5000, 7500, 10,000 ppm. For further study, the water content is from 0, 20, 40 and 60%. Each conditions are tested for 3 times to get the average results.

3 Results and Discussion

3.1 Chemical Screening

Based on the type of chemicals, the solvent inhibitor (n-heptane, n-hexane) and the crystal modifiers (PMAO and MEA), the experiment is conducted at different concentration and presents in Figs. 1 and 2 for solvent inhibitor and crystal modifier, respectively.

Solvent Inhibitor. From Fig. 1, it is presents that the original oil has pour point at 36.3 °C. When n-heptane and n-hexane are applied from 5 to 20%, the pour points are much lower to 21.0 °C and 21.3 °C, respectively, for 0% water cut. However, at 20% water cut, pour point of original oil becomes increased to 37.3 °C and the pour point will decrease as solvent inhibitors are applied up to 21.7 °C for n-heptane and 21.9 °C for n-hexane. Also, n-heptane and n-hexane can effectively lower the pour point of crude oil up to 42.15% and 41.32%, respectively. The reason for the decline of pour point using solvent inhibitor is that they have similar structure to hydrocarbon in crude oil. Also, the light hydrocarbon type will act as solvent to the wax particle [10].

Therefore, n-heptane has the rather better performance to reduce pour point of crude oil. Therefore, n-heptane will be selected to study further.

Crystal Modifier. The crystal modifiers of PMAO and MEA for pour point reduction is presented in Fig. 2. The concentration used is in ppm level. It is much lower than solvent type because of being crystal modifier type inhibitor. In both original crude oil (0% water cut) and 20% water volume fraction cases, the concentration varying from 2500 to 10,000 ppm. For chemical screening, it is clear that these chemicals have less effect on pour point reduction at any varying conditions. The reason is that the crystal modifier helps to present the formation of wax by

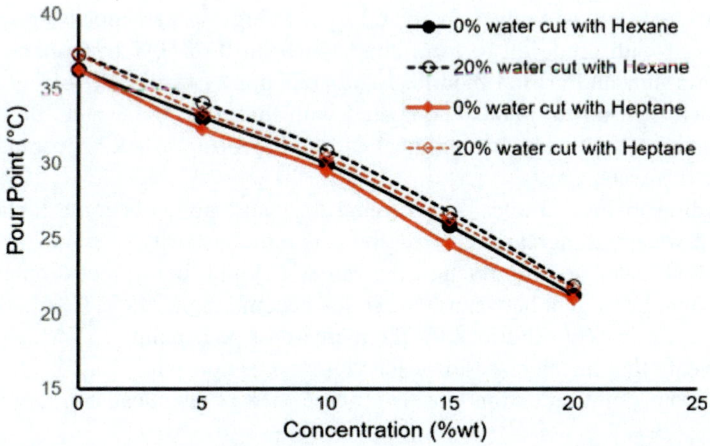

Fig. 1 Result of concentration of solvent inhibitor on pour point

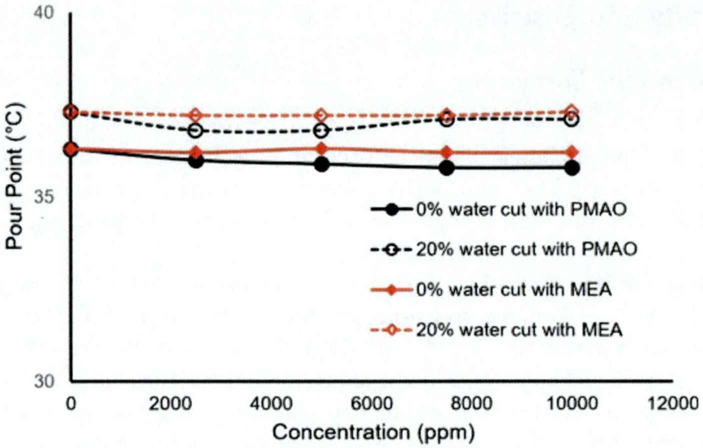

Fig. 2 Result of concentration of crystal modifier type inhibitor on pour point

arranging the crystal for wax deposition. Therefore, it can be used for different mechanism for wax deposition prevention [11].

In addition, from Fig. 2, PMAO can provide the lower pour point reduction compared to MEA. However, PMAO is in the form of solid and is more difficult to use in this study. Moreover, MEA is in the form of liquid, and the molecule is smaller. Also, it is interesting to use and study for wax prevention. Therefore, this chemical will be selected for further study.

3.2 *Effect of Emulsion on Pour Point Reduction*

The emulsion for this study is prepared by varying the percentage or amount of water mixed with crude oil to form emulsion from 0 to 60% by volume for both solvent inhibitor and crystal modifier. The pour point of oil with solvent inhibitor can be increased up to 30.08% compared with that of at 0% water cut. Also, the pour point of oil with crystal modifiers can be higher up to 8.29% compared with that of at 0% water cut.

According to Figs. 3 and 4, it is found that pour points become higher as the amount of water cut increases. The reason is that the viscosity of emulsion becomes higher when water volume fraction increases [12] and the droplet-droplet interaction and interfacial area between two phases become higher [9, 13, 14]. Therefore, viscosity is one of contributions for the increase of pour point [15]. Moreover, this study reveals that an increase in water volume fraction will result in diminishing effect on pour point reduction of solvent type of wax chemical inhibitor.

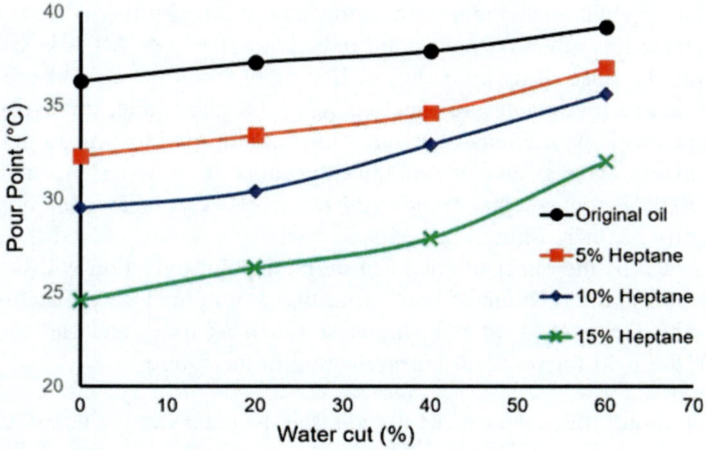

Fig. 3 Emulsion effect on pour point of emulsion using n-Heptane

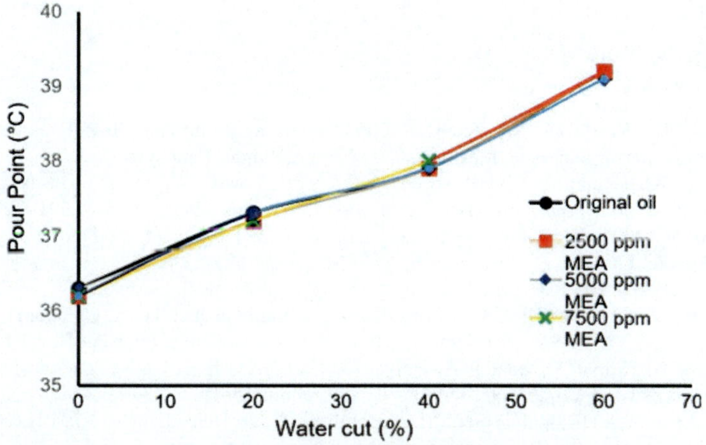

Fig. 4 Emulsion effect on pour point of emulsion using MEA

4 Conclusion

The effective method for wax prevention is chemical method because of continuous operation and an ease of use. Pour point reduction is one of the mechanisms to reduce wax deposition. Moreover, water produced along with oil can be mixed with oil to form emulsion. In this study, the potential chemicals, n-heptane, n-hexane as a solvent inhibitor and PMAO and MEA as a crystal modifier for pour point reduction are tested and selected. The results show that solvent inhibitor especially n-heptane provides the better result for pour point reduction up to 42.15%

compared to original oil and it is selected to study further. Furthermore, both crystal modifiers have less effect on pour point reduction. However, MEA is selected for further study because of its properties of liquid phase and an ease of use.

In addition to investigate the effect emulsion on pour point, the experiment of selected chemicals with emulsion varying the amount of water cut from 0 to 60% has been made. The results present that the amount of water in emulsion can increase 30.08% for n-heptane and 8.29% for MEA compared to original oil. Consequently, an increasing of percentage of water cut is one of factors to increase pour point, whilst the effect of chemical on pour point reduction will be reduced. From this study, the problem of wax deposition from crude oil production can be lowered with the pour point reducing agent such as n-heptane and this can be applied in the field for practical oil production in the future.

Acknowledgements The authors would like to acknowledge the Defence Energy Department, Ministry of Defence at Fang Oilfield, Chiang Mai, Thailand for the samples of oil. Also, the authors would like to thank Chevron Thailand Exploration and Production Ltd., for financial support of this project.

References

1. Totten GE, Westbrook SR, Shah R (2003) Fuels and lubricants handbook: technology, properties, performance, and testing. ASTM International, Pennsylvania
2. Pedesen KS, Ronningsen HP (2003) Influence of wax inhibitors on wax appearance temperature, pour point, and viscosity of waxy crude oils. Energ Fuel 17:321–328
3. Fanchi JR (2007) Petroleum engineering handbook, vol 1. Society of Petroleum Engineers, Richardson, TX
4. Allen TO, Roberts AP (1989) Production operations: well completions, workover, and stimulation, vol 2 ed. Oil & Gas Consultants International Inc, Tulsa, Oklahoma
5. Theyab MA, Yahya SY (2018) Introduction to wax deposition. Int J Pet Res 2(1):126–131
6. Banerjee S, Kumar S, Mandal A, Naiya TK (2017) Design of novel chemical solvent for treatment of waxy crude. Int J Oil Gas Coal Technol 15(4):363–379
7. Jafari Ansaroudi HR, Vafaie-Sefti M, Masoudi Sh, Jafari Behbahani T, Jafari H (2013) Study of the morphology of wax crystals in the presence of ethylene-co-vinyl acetate copolymer. J Pet Sci 31(6):643–651
8. Kokal SL, Aramco S (2006) Crude oil emulsions. Society of Petroleum Engineers, Richardson, TX
9. Farah MA, Oliveira RC, Caldas JN, Rajagopal K (2005) Viscosity of water-in-oil emulsions: variation with temperature and water volume fraction. J Petrol Sci Eng 48:169–184
10. Kang PS, Lee DG, Lim JS (2014) Status of wax mitigation technologies in offshore oil production. In: Proc. 24th International Ocean and Polar Engineering Conference, Busan, Korea, June, pp 31–38
11. Min HM, Maneeintr K (2019) Investigation of wax deposition prevention by using n-heptane for sustainable energy production from Fang oilfield. IOP Conf Ser Earth Environ Sci 268
12. Maneeintr K, Sasaki K, Sugai Y (2013) Investigation of the effects of parameters on viscosities and correlation of heavy oil and stability of its emulsion. J Japan Inst Energ 92:900–904
13. Wutthisirisart S (2014) Analysis of the effects of viscosity and shear rate on light oil and its emulsion from an oilfield in Thailand, Master Thesis, Chulalongkorn University

14. Juntrarasakul O, Maneeintr K (2014) Assessment of CO_2 geological storage potential in a depleted oil field in the North of Thailand. IOP Conf Ser Earth Environ Sci 175–179
15. Emmanuel O, Egbogah Jack T (1990) An improved temperature viscosity correlation for crude oil system. J Pet Sci Technol 5:197–200

Research on the Defects in Graphene: Types and Effects on Mechanical and Thermal Properties

Huishu Wang

1 Introduction

Graphene, consisting of only one layer of graphite, is the first two-dimensional material that has been characterized. In graphene, the carbon atoms are bonded in the honeycomb lattice structure. Figure 1 presents that graphene is the elementary unit of the other allotropes of carbon [1]. In the 1930s, Peierls and Landau presumed that thermal fluctuations would cause the instability of two-dimensional crystals at finite temperature and 2D crystals could not exist in reality. The argument was supported by Mermin's theorem that long-range ordering crystal formation was impossible in 2D atomic structures. Varies experimental investigation also proved thin films would become unstable with reduced thickness. However, Geim and Novoselov used tape to successfully isolate single-layer graphene from thin carbon films in 2004.

Thanks to its special monatomic structure, graphene has been considered as one of the most prospective materials, having outstanding and unique properties. In each carbon atom, there are four outer-shell electrons, three of them contribute to forming σ-bonds with three other carbon atoms through sp2 hybrid orbitals [2]. These strong σ-bonds restrict in-plane deformation and could explain the excellent mechanical properties of graphene. The remaining free-moving electron perpendicular to the plane could benefit the electronic properties of graphene. Graphene also has superior thermal, magnetic properties and high specific surface area. Phonons (or lattice vibrations) are responsible for the high thermal conductivity in graphene [3]. Graphene has been considered the strongest material and has the highest thermal conductivity in the world; therefore, it becomes an ideal candidate in graphene-based nanocomposites and thermal management. However, 2D materials are difficult to occur naturally and can only be obtained by exfoliation from natural materials or synthesis technology. No matter which method is adopted,

H. Wang (✉)
University of Bristol, Bristol BS8 1TH, UK

R. K. Agarwal (ed.), *Recent Advances in Manufacturing Engineering and Processes*, Lecture Notes in Mechanical Engineering,
https://doi.org/10.1007/978-981-16-3934-0_19

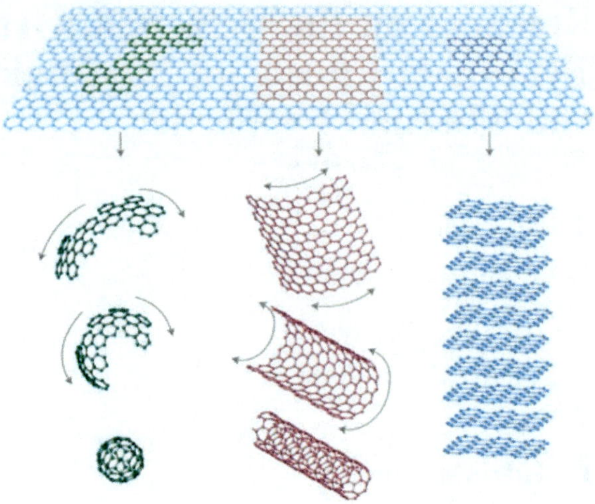

Fig. 1 How does graphene related to other graphitic forms such as 0D buckyballs, nanotubes and graphite (from left to right) [1]

different types of defects will be introduced in graphene inevitability and the lattice imperfections do have a significant impact on the properties of pristine graphene [4]. Besides, methods such as irradiation and chemical treatment could be utilized to introduce defects into graphene purposely to tailor the regional properties of graphene, which makes the graphene accomplish different functionalities [5]. For example, defected graphene has been used in thermoelectricity, thermal insulation and thermal rectification [6]. Hence, trying to understand, the influence of defects is valuable. This paper will introduce different categories of defects within graphene and illustrate a review of defects study to show the impact of defects on the mechanical and thermal properties of graphene.

2 Types of Defects in Graphene

The perfect graphene is two dimensional and contains only a regular hexagon lattice structure. However, graphene is not perfectly flat in reality. Also, due to the synthesis process, different types of defects will be introduced. Generally, there are two categories of defects, intrinsic defects and extrinsic defects.

2.1 Intrinsic Defects

Intrinsic defects refer to the perturbed crystalline order. There are five types of intrinsic defects: Stone-Wales defects, single vacancies, multiple vacancies, line defects and carbon adatoms. Line defects are one-dimensional defects, while the others are zero-dimensional defects [4].

Within the Stone-Wales defect, there are no addition and removal of carbon atoms and it could be simply developed as a result of an in-plane 90 °C rotation of one of the C–C bonds. In this way, graphene lattice is rearranged and adjacent pairs of pentagons and heptagons are formed, which is shown in Fig. 2a. This kind of defect could be ignored at temperatures below 1000 °C because the equilibrium concentration is extraordinarily small due to the high absolute formation energy which is about 5 eV. However, Stone-Wales defect could be deliberately introduced by particle irradiation with ions or electrons.

The other types of defects involve added or removed atoms. Figure 2b indicates the simplest form of defects, single vacancy, containing a five-membered and a nine-membered ring [4]. Missing one carbon atom results in three dangling bonds which connected to it previously. Because Jahn-Teller distortion tries to minimize the total energy, therefore two dangling bonds are connected and only one dangling bond remains. The absolute formation energy of single vacancies is higher than Stone-Wales defects with the experimental value of approximately 7.0 eV [7].

As for the multiple vacancies, there are various lattice structures. Figure 2c–e shows three types of double vacancy defect structures. Double vacancies are created by losing one more atom on the basis of single vacancies. Figure 2c is the easiest structure to imagine, and the formation energy is around 8 eV [8]. While the construction of V2 (555–777) defect shown in Fig. 2d is more possible due to lower formation energy is required [4].

Line defects could be considered as two domains of diverse lattice orientations with the tilt axis being separated by tilt boundaries [4]. The introduction of this type of defect is always due to the random growth of graphene within chemical vapor deposition [9]. CVD is one of the synthesis approaches to obtain graphene. Graphene could grow from C-containing gases on catalytic metal surfaces. One example of line defects is the growth of two domains on a Ni surface as shown in

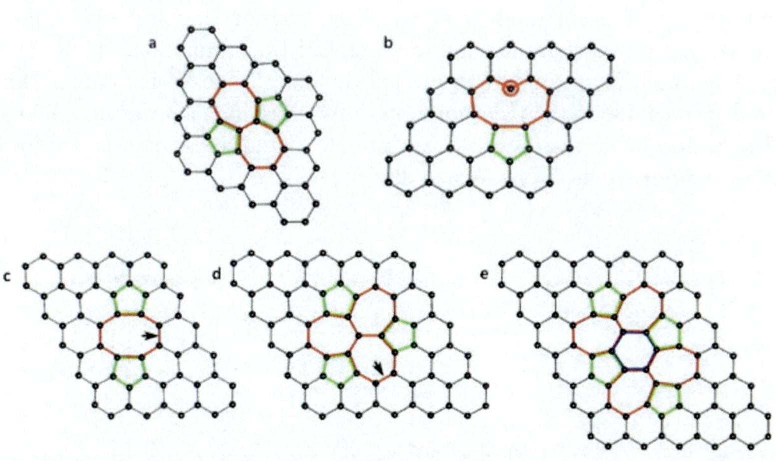

Fig. 2 Atomic structures of **a** Stone-Wale defect, **b** single vacancy and **c–f** double vacancies obtained from DFT calculation [4]

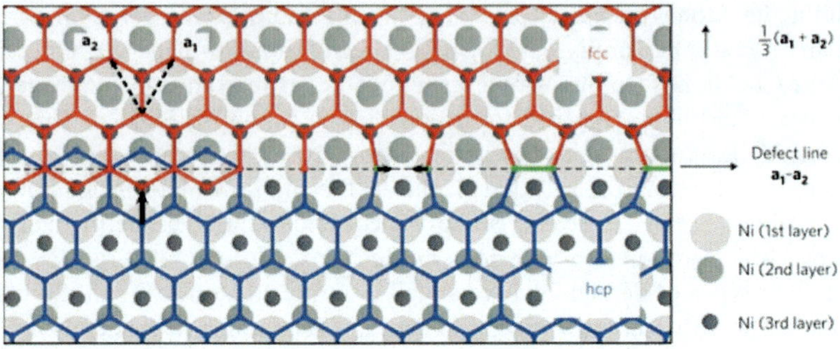

Fig. 3 Dislocation defect consisting of octagons and pentagons when two half-lattices grow on Ni substrate [10]

Fig. 3 [10]. The domain boundary is reconstructed and a periodic structure consisting of pairs of pentagons separated by octagons.

Carbon adatoms refer to the defects when the removal carbon atoms generated from the vacancies migrate are separated from the original carbon ring before migrating on the surface of the graphene [9]. These carbon atoms could form new bonds with other atoms. The missing carbon atoms might indeed fill the vacancies as they migrate. While if the missing atoms migrate to the region containing no vacancies, then changes of the hybridization of the carbon atoms in the region would take place and new defects would be introduced. The additional atoms would use the third dimension instead of straining the local structure in two dimensions [4]. Figure 4 presents three configurations of carbon adatoms. The adatom on top of a carbon–carbon bond in the bridge configuration shown in Fig. 4a is the energetically preferred configuration. Adatoms could migrate easily over graphene surfaces, which makes other configurations possible and leads to a complex behavior. Figure 4b occurs in the case when the adatoms migrate through the lattice by an exchange mechanism, leading to a dumbbell configuration which has higher energy than the bridge configuration. The inverse Stone-Wales defect shown in Fig. 4c is formed due to the incorporation of two missing adatoms into the network of sp2-hybridized carbon atoms [4]. Such defect could be ignored as the formation energy of is higher than Stone-Wales defect.

Fig. 4 Three types of carbon adatoms [4]

2.2 *Extrinsic Defects*

The extrinsic defects refer to the introduction of foreign atoms [4]. The extrinsic defects could be subdivided into two categories which are foreign adatoms and substitutional impurities.

Foreign adatoms, compared to carbon adatoms, refer to the situation when noncarbon atoms or functional groups are introduced to the exterior of graphene. Foreign adatoms are inevitably introduced due to chemical treatment. During chemical vapor deposition, metal atoms could be bonded with the adjacent carbon atoms [9]. The effect of this kind of defects on graphene properties is related to the bonding. Only physisorption as a result of van der Waals interaction occurs for the weak bond, but chemisorption could occur due to covalent bonding if there is a stronger interaction between the foreign atom and adjacent carbon atoms [4]. In addition to the introduction of metal atoms, oxygen atoms or oxygen-containing functional groups such as hydroxyl groups or carboxyl groups could have greater influence on the properties of graphene. Such defects are formed as a result of strong oxidation being used during the Hummers method which is a kind of graphene preparation method [9].

The introduced foreign atoms could also replace one or two carbon atoms in graphene, leading to substitutional impurities. Boron and nitrogen are considerable candidates to replace carbon atoms because they have similar lattice parameters. The electronic structure of graphene can be altered by deliberately introducing boron and nitrogen, which enables the evolution of applications in electronics and optics with distinct properties [11].

3 Effect of Defect on Mechanical Properties

The mechanical properties of single-layer graphene are firstly measured by nanoindentation in an AFM. The results show a Young's modulus of 1 TPA and breaking strength of graphene and intrinsic strength are 42 N/m and 130 GPa, respectively, [12]. Graphene is considered the strongest material because of these impressive mechanical properties.

Studies of properties of defective graphene have been carried out mostly by molecular dynamic simulation. The potential functions chosen have a significant impact on the accuracy of results predicted [13]. But the effects of graphene on properties after introducing defects have a similar variation trend. Feng Hao et al. used adaptive intermolecular reactive empirical bond order potential functions to simulate the interactions between carbon atoms. Results indicated that there was a linear dependence between Young's modulus and monovacancy and a much smooth relationship between Young's modulus and SW defects, but the effect of both defects was gentle [14]. Also, non-equilibrium molecular dynamics (NEMD) based on optimized Tersoff potential was used by Mortazavi and Ahzi [13] to study defective graphene. Their results were similar to Feng Hao et al. and they found that increasing concentration of point vacancy, Stone-Wales and bivacancy defects led to the decrease in elastic modulus. From the study of dislocation in nanotube with computer simulation [15], it is believed that the line effect contributes to plastic deformation of graphene ribbon [4]. Furthermore, based on the second-generation REBO potential, atomic finite element (AFM) had been used to study complex vacancy and topological defects. And the results indicated that the relationship between the topological defect and tensile strength was complex, also suggested that AFEM was a potential tool to study topological optimization [16].

Experiments have also been conducted to report the mechanical properties of defective graphene. The impact of sp3 type defects (graphene is connected to an oxygen adatom) and vacancy defect on elastic modulus and breaking strengthen was studied by Zandiatashbar et al. using AFM nanoindentation. They indicated that sp3 type defects had a really small impact on elastic modulus and breaking strength is only approximately 14% smaller than pristine graphene. However, there was a significant drop in both elastic modulus and breaking strength within the vacancy defect regime [17].

4 Effect of Defect Type on Thermal Properties

Thermal conductivity of single-layer graphene at room temperature has been measured to be approximately 4840–5300 W/mK through experimental investigation using confocal micro-Raman spectroscopy [18]. While thermal conductivities calculated from molecular dynamic simulation by using different potential functions vary a lot and are far below the experimental results [19].

Mortazavi et al. also obtained results of the relationship between concentration of vacancies and thermal conductivity. They observed 50% reduction in thermal conductivity as a result of 0.25% concentration of vacancy defects, and thermal conduction was more sensitive to defect than elastic modulus [13], which maps with the results obtained by Hao et al. [14]. The sensitivity between increase of vacancy concentration and reducing thermal conductivity was due to a transition from propagating to diffusive mode as phonons were scattered by defects [14]. Xin Wu et al. also conducted the research on the effect of SW defect and vacancies on thermal conductivity. They adopted GPU-accelerated molecular dynamics software and EMD method with Tersoff potential. They found that the SW defect had less effect compared to vacancies and the reducing rate tended to be more smooth as increasing defects concentration [19].

There are also studies on the influence of extrinsic defects on thermal conductivity. Chien et al. [20] investigated the influence on thermal conductivity due to the introduction of hydrogen functionalization using non-equilibrium molecular dynamics simulation. They simulated the bonding interaction between C–C and C–H choosing second-generation reactive empirical bond order potential and found that thermal conductivity reduced by 40% due to introducing hydrogen into 2.5% of graphene carbon atoms. They also explained the decline of thermal conductivity as a result of the limit in the phonon mean free path.

5 Conclusion

In this paper, different types of defects are introduced following by a review of the impact of defective graphene on mechanical and thermal properties. In summary, sp3 type defect has a relatively small effect on elastic modulus and strength of graphene, which means that reinforcing abilities of graphene could maintain when bonded to polymer matrices. The insensitivity to sp3 type defect makes graphene having a promising prospect in nanocomposite materials. While vacancies do have a significant influence on the mechanical properties of pristine graphene and SW defects have less impact. Besides, dislocations need to be considered in the plastic deformation of graphene ribbons. Moreover, promising thermal conductivity of pristine graphene is more vulnerable than mechanical properties. Because of difficulties to control defects and simulate the real interaction between atoms, it is hard to derive the quantitative relationship between defects and properties of graphene. It is crucial to further understand the impact of defects on the properties of graphene and to develop the innovative application of defective graphene.

References

1. Geim AK, Novoselov KS (2007) The rise of graphene. Nat Mater 6(3):183–191. https://doi.org/10.1038/nmat1849
2. Yang G, Li L, Lee WB, Ng MC (2018) Structure of graphene and its disorders: a review. Sci Technol Adv Mater 19(1):613–648. https://doi.org/10.1080/14686996.2018.1494493
3. Tiwari SK, Sahoo S, Wang N, Huczko A (2020) Graphene research and their outputs: status and prospect. J Sci Adv Mater Devices 5(1):10–29. https://doi.org/10.1016/j.jsamd.2020.01.006
4. Banhart F, Kotakoski J, Krasheninnikov AV (2011) Structural defects in graphene. ACS Nano 5(1):26–41. https://doi.org/10.1021/nn102598m
5. Liu L, Qing M, Wang Y, Chen S (2015) Defects in graphene: generation, healing, and their effects on the properties of graphene: a review. J Mater Sci Technol 31(6):599–606. https://doi.org/10.1016/j.jmst.2014.11.019
6. Pyun KR, Ko SH (2019) Graphene as a material for energy generation and control: recent progress in the control of graphene thermal conductivity by graphene defect engineering. Mater Today Energy 12:431–442. https://doi.org/10.1016/j.mtener.2019.04.008
7. Thrower PA, Mayer RM (1978) Point defects and self-diffusion in graphite. Phys Status Solidi 47(1):11–37. https://doi.org/10.1002/pssa.2210470102
8. Krasheninnikov AV, Lehtinen PO, Foster AS, Nieminen RM (2006) Bending the rules: contrasting vacancy energetics and migration in graphite and carbon nanotubes. Chem Phys Lett 418(1–3):132–136. https://doi.org/10.1016/j.cplett.2005.10.106
9. Tian W, Li W, Yu W, Liu X (2017) A review on lattice defects in graphene: types generation effects and regulation. Micromachines 8(5). https://doi.org/10.3390/mi8050163

10. Lahiri J, Lin Y, Bozkurt P, Oleynik II, Batzill M (2010) An extended defect in graphene as a metallic wire. Nat Nanotechnol 5(5):326–329. https://doi.org/10.1038/nnano.2010.53
11. Ci L et al (2010) Atomic layers of hybridized boron nitride and graphene domains. Nat Mater 9(5):430–435. https://doi.org/10.1038/nmat2711
12. Lee C, Wei X, Kysar JW, Hone J (2008) Measurement of the elastic properties and intrinsic strength of monolayer grapheme. Science(80-) 321(5887):385–388. https://doi.org/10.1126/science.1157996
13. Mortazavi B, Ahzi S (2013) Thermal conductivity and tensile response of defective graphene: a molecular dynamics study. Carbon N.Y. 63:460–470. https://doi.org/10.1016/j.carbon.2013.07.017
14. Hao F, Fang D, Xu Z (2011) Mechanical and thermal transport properties of graphene with defects. Appl Phys Lett 99(4):041901. https://doi.org/10.1063/1.3615290
15. Ding F, Jiao K, Wu M, Yakobson BI (2007) Pseudoclimb and dislocation dynamics in superplastic nanotubes. Phys Rev Lett 98(7):075503. https://doi.org/10.1103/PhysRevLett.98.075503
16. Damasceno DA, Rajapakse RKND, Mesquita E, Pavanello R (2020) Atomistic simulation of tensile strength properties of graphene with complex vacancy and topological defects. Acta Mech 231(8):3387–3404. https://doi.org/10.1007/s00707-020-02715-6
17. Zandiatashbar A et al (2014) Effect of defects on the intrinsic strength and stiffness of grapheme. Nat Commun 5. https://doi.org/10.1038/ncomms4186
18. Balandin AA et al (2008) Superior thermal conductivity of single-layer graphene. Nano Lett 8 (3):902–907. https://doi.org/10.1021/nl0731872
19. Wu X, Han Q (2020) Thermal conductivity of defective graphene: an efficient molecular dynamics study based on graphics processing units. Nanotechnology 31(21). https://doi.org/10.1088/1361-6528/ab73bc
20. Chien S-K, Yang Y-T, Chen C-K (2011) Influence of hydrogen functionalization on thermal conductivity of graphene: nonequilibrium molecular dynamics simulations. Appl Phys Lett 98 (3):033107. https://doi.org/10.1063/1.3543622

PDMS-Coated *Rosa Centifolia* Flower: Characterization and Their Stability in Toner Solutions

Dina Febriani, Indah K. Sulistiyorini, Yoki Yulizar, Dewangga Oky Bagus Apriandanu, Rizki M. Surya, Andhina R. Satriani, Ariffinisa L. Widyaningtyas, and Cheryl Ariela

1 Introduction

Rose flower (*Rosa centifolia*) is an ornamental plant, which has been widely used in the cosmetic industry due to its easy preparation, low cost, non-toxic, environmentally friendly and fully biodegradable [1]. *Rosa centifolia* has a tonic and an astringent effect of tightening the skin, shrinking pores and reducing inflammation. Rose petals are useful for rejuvenating the skin, anti-depressant, antitussive, inhibiting vasoconstriction and antibacterial [2].

To date, rose is utilized in cosmetic products, such as toner, to add aesthetic value and to enhance a product's natural image. The compounds, such as phenyl ethanol (43%), geranyl acetate (15.6%), geraniol (10.5%), linalool (6.9%), benzyl alcohol (3.3%), benzaldehyde (1.5%), nerol (5–10%), citronellyl acetate (0.3%) were reported. Also, the additional substance were investigated such as tannin compounds, oligomeric proanthocyanidins, sacaine, mineral salts, malic acid and tartaric acid salts, pectin (11%), riboflavin, sugar, laxative glycosides (multiflorine A and B) [3].

Natural ingredients without special treatment can be degraded more quickly in the toner solution due to the surface hydrophobic wax on the rose. There is still an interaction between the air and the surface of the flower. Therefore, the rose sheet needs to be modified to obtain a superhydrophobic surface, which is more durable, stable and not damaged in the toner solution. The surface of rose is superhydrophobic if it has a water contact angle of less than 150° [4]. Superhydrophobic surfaces have been developed through several methods, such as sol–gel, lithogra-

D. Febriani · I. K. Sulistiyorini · Y. Yulizar (✉) · D. O. B. Apriandanu · R. M. Surya
Department of Chemistry, Faculty of Mathematics and Natural Science, Universitas Indonesia, Kota Depok, Indonesia
e-mail: yokiy@sci.ui.ac.id

A. R. Satriani · A. L. Widyaningtyas · C. Ariela
PT Nose Herbal Indo, Sunter Agung, Tanjung Priok, Jakarta Utara, Indonesia

R. K. Agarwal (ed.), *Recent Advances in Manufacturing Engineering and Processes*, Lecture Notes in Mechanical Engineering,
https://doi.org/10.1007/978-981-16-3934-0_20

phy, casting, electrospinning, chemical vapour deposition (CVD), dip-coating and templating. However, among these methods, the dip-coating is an inexpensive and easy to form superhydrophobic surfaces in a wide range of applications [5]. This method immerses the substrate in a solution at a certain time and pulls it vertically at a constant speed [6]. The superhydrophobic surface can be obtained by making a rough surface, then modifying its surface with a low surface energy material, for instance, polydimethylsiloxane (PDMS) [7].

In recent years, polydimethylsiloxane (PDMS) has become a suitable candidate for the fabrication of superhydrophobic surfaces for oil-water separation due to its low surface energy, high biocompatibility, strong adhesion, ease of processing, good chemicals and high thermal stability [8].

In this study, rose flower (*Rosa centifolia*) was modified its surface through a dip-coating method by using PDMS as surface coating materials to form a superhydrophobic property on the flower's surface. Eventually, the physical stability of rose increases in the toner solution.

2 Materials and Methods

2.1 Materials

Rosa centifolia and toner solution were obtained from PT Nose Herbal Indo, Jakarta, Indonesia. The chemicals used in this research, such as toluene, polydimethylsiloxane (PDMS) and ethanol were purchased from Merck, Germany. Aquadest was used without purification.

2.2 Preparation of Rosa Centifolia

The pre-treated flower was performed through a method as reported by Indriyani et al. [9]. *Rosa centifolia* was cleaned and dried at room temperature. Then, the sample was dried for seven days to remove the moisture content in roses.

2.3 Preparation of PDMS Solution

PDMS was added into 25 mL of toluene (0.5%, w/v), then stirred for 60 min at 80°C. After that, the solution was cooled to room temperature. 10 mL of ethanol was added into the mixture, then stirred back until dissolving completely.

2.4 Dip-Coating of Rosa Centifolia

Rosa centifolia were immersed in a PDMS coating solution for 30 min, then pulled vertically at a constant speed and dried for 5 min at room temperature. This procedure was repeated nine times to increase the adhesion of the coating of the substrate. Then, the samples were dried for an hour at room temperature, then dried at 40 °C for 30 min. Finally, the result was rinsed and stored for 24 h by observing its colour change.

2.5 Characterization of Rosa Centifolia Surface

The Attenuated Total Reflection-Fourier Transform Infra-Red (ATR-FTIR) spectrum was performed via a spectrometer (Thermo Scientific Nicolet 6700) in the 400–4000 cm^{-1} region to identify the functional groups of the modified surface of rose. The static contact angle (CA) was measured by putting 5 μL of water droplets on each sample surface. The measurements were taken at three different substrate regions at room temperature. Each CA was obtained by calculating the average of the measurements using ImageJ software. UV–Visible Spectrophotometer (Shimadzu 2600) was used to monitor the absorption spectra of PDMS in toner solution.

2.6 Stability of Rosa Centifolia in Toner Solution

In the cycling test method, uncoated and coated roses were put into a toner solution and stored at 40 °C for 24 h, then dried at 40 ± 2 °C for 24 h (one cycle). The changes of toner solution and rose flower were observed every two days.

3 Results and Discussions

PDMS-coated roses were characterized using ATR-FTIR to determine the interaction of functional groups on the surface of roses. Figure 1 shows the ATR-FTIR spectra of rose before and after surface modification with PDMS. Based on the ATR-FTIR analysis, the spectra show the O–H stretching at 3346.88 cm^{-1}, C–H sp^3 asymmetric and symmetric with stretching vibration at 2917.29 and 2849.67 cm^{-1}, C–O stretching at 1728.71 cm^{-1}, C–C stretching at 1609.49 cm^{-1}, O–H bending at 1461.93 cm^{-1}, and C–O stretching at 1179.10 and 1011.08 cm^{-1}. Also, the spectra of PDMS-coated roses show the specific peak of PDMS at 1017.50 and 1103.29 cm^{-1}, indicating the presence of Si–O–Si bonds [10].

In addition, there is a peak of absorption at 1640.57 cm^{-1}, which corresponds to Si–OH bonds [11]. These results demonstrate that the rose has successfully modified its surface with PDMS.

Figure 2 shows the water contact angle dripping on the surface of rose. The uncoated rose exhibited a water contact angle of 138.4° (Fig. 2a). It can be seen that roses have hydrophobic properties with a low level of wetness to water. In order to increase its ability to repel water, the rose was coated with PDMS. Figure 2b shows that PDMS-coated rose presented the water contact angle of 150°. The presence of PDMS as coating material decreases surface energy and increases the surface roughness of roses. The increase of the water contact angle illustrates that the surface of rose was successfully made as a superhydrophobic layer by PDMS.

Figure 3 shows the stability of rose in toner solutions observed by UV–Vis spectrometry. The toner solution shows the absorbance of 0.423 at a wavelength of 246 nm. The UV–Vis spectra display that there was an increase in absorbance, indicating an unstable rose flower surface in the toner solution [12]. The wavelength change indicates that the rose has a potential for degradation. The recorded maximum wavelength to observe the stability of rose is tabulated in Table 1. Moreover, the instability of rose was observed from its colour as shown in Fig. 4. The colour of the rose was influenced by the existence of anthocyanin compounds in flowers. Uncoated roses displayed a colour change from red to brown due to diffusion through the flower pores. Therefore, the dye from anthocyanin easily degrades in the toner solution and affects its colour. However, the absorbance of PDMS-coated rose was insignificantly changed as shown in Fig. 3. It is noteworthy that the coated rose was more stable in toner solution than the uncoated rose.

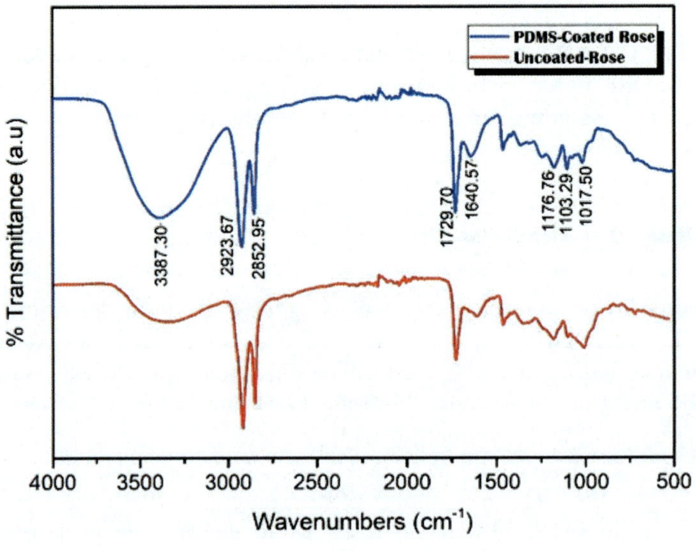

Fig. 1 ATR-FTIR spectra of uncoated rose and PDMS-coated rose

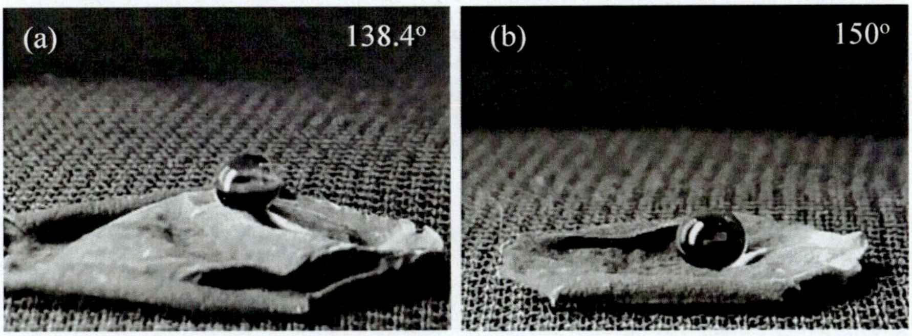

Fig. 2 Water contact angle on the surface of **a** uncoated rose and **b** PDMS-coated rose

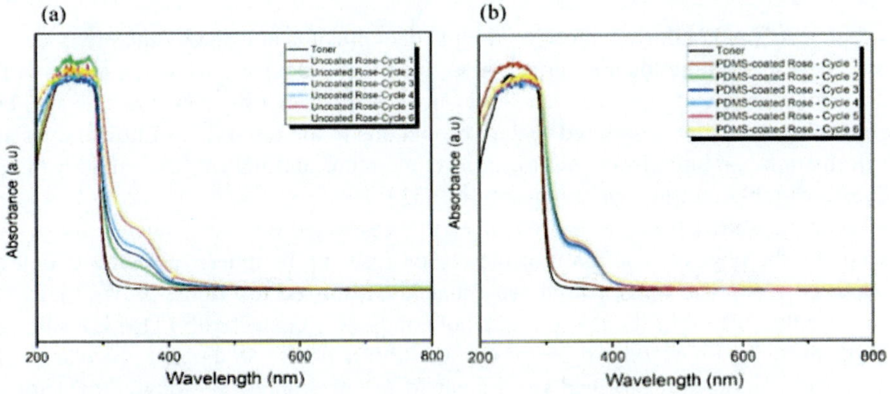

Fig. 3 UV–Vis absorption spectra of **a** uncoated rose and **b** PDMS-coated rose

Table 1 The stability of rose at various cycles

Cycle	Uncoated rose		PDMS-coated rose	
	λ max (nm)	Absorbance (A)	λ max (nm)	Absorbance (A)
1	260	0.450	258	0.447
2	261	0.477	287	0.447
3	278	0.484	287	0.447
4	272	0.492	262	0.448
5	279	0.515	249	0.448
6	258	0.519	269	0.449

Figure 4 shows the stability results of uncoated roses for 6 cycles. The degree of acidity (pH) of the toner before adding the roses was 4.5. Once the roses were added, the degree of acidity decreased up to cycle 6, which were found to be 4.3;

Fig. 4 Cycling test of uncoated roses in toner solution at **a** cycle 0, **b** cycle 1, **c** cycle 2, **d** cycle 3, **e** cycle 4, **f** cycle 5, **g** cycle 6

4.3; 4.2; 4.2; 4.0; 4.0, respectively. The pH decrement is affected by the presence of dyes degraded in the toner solution. Also, the toner colour was changed along with the enhancement of the cycle as shown in Fig. 4. It can be seen that the flower colour converted from dark red to dark brown due to the instability of flower dyes at extreme temperatures from low to high temperature and pH of 4.5. Anthocyanins dyes are stable at a low pH range of 1–2 [13].

Figure 5 shows the stability results of PDMS-coated rose for 6 cycles. It can be seen that the physical condition of the coated roses in the toner was better than the uncoated rose. The toner colour was changed from red to orange-brown. This is because the surface of the flowers was not completely coated with PDMS, resulting in the diffusion on of toner solution into the flower pores. As a result, the clarity of the toner solution transformed from clear to bright orange. The degree of acidity (pH) of the toner before adding the roses was 4.5. Once the roses were added, the degree of acidity decreased was stable at 4.5 up to cycle 6. It is important to note that the addition of PDMS did not possess any impact on the degree of acidity of the toner.

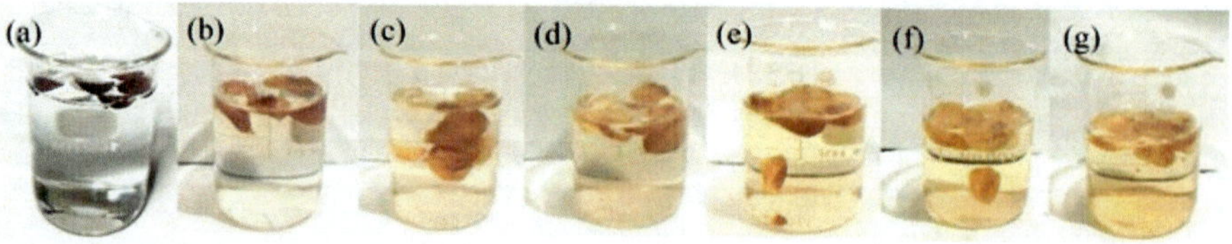

Fig. 5 Cycling test of PDMS-coated roses in toner solution at **a** cycle 0, **b** cycle 1, **c** cycle 2, **d** cycle 3, **e** cycle 4, **f** cycle 5, **g** cycle 6

4 Conclusions

The surface of rose (*Rosa centifolia*) was successfully modified with PDMS through the dip-coating method. According to ATR-FTIR analysis, the presence of typical Si–O–Si bonds appeared at 1017.08 and 1095.26 cm^{-1} derived from PDMS. Roses successfully possessed superhydrophobic surface characteristics after a modification with PDMS and generated a high-water contact angle of 150°. Notably, the PDMS-coated roses were more stable in toner solution than the uncoated rose with an insignificant discoloration of the toner and a stable pH at 4.5.

References

1. Thambidurai M, Muthukumarasamy N, Velauthapillai D, Lee C (2014) Rosa centifolia sensitized ZnO nanorods for photoelectrochemical solar cell applications. Sol Energy 106:143–150
2. Ashokirao T, Kiran D, Suryawanshi M (2017) A brief review on therapeutic effects of —'Ornamental Plant' rose. Int J Ayurveda Pharma Res 5(12):46–52
3. Jitendra J, Vineeta T, Ashok K, Brijesh K, Singh P (2012) Rosa centifolia: plant review. Int J Res Pharm Chem 2(3):794–796
4. Parvate S, Dixit P, Chattopadhyay S (2020) Superhydrophobic surfaces: insights from theory and experiment. J Phys Chem B 124(8):1323–1360
5. Janeiro P, Brett O, Maria A (2004) Catechin electrochemical oxidation mechanisms. Anal Chim Acta 518(1–2):109–115
6. Gurav AB et al (2015) Superhydrophobic coatings prepared from methyl-modified silica particles using simple dip-coating method. Ceram Int 41(2):3017–3023
7. Li G, Xu S (2015) Small diameter microchannel of PDMS and complex three-dimensional microchannel network. Mater Des 81(May):82–86
8. Barthwal S, Barthwal S, Singh B, Singh NB (2020) Multifunctional and fluorine-free superhydrophobic composite coating based on PDMS modified MWCNTs/ZnO with self-cleaning, oil-water separation, and flame retardant properties. Colloids Surf A Physicochem Eng Asp 597:124776
9. Indriyani A, Yulizar Y, Yunarti RT, Apriandanu DOB, Surya RM (2021) One-pot green fabrication of $BiFeO_3$ nanoparticles via *Abelmoschus esculentus* L. leaves extracts for photocatalytic dye degradation. Appl Surf Sci 563:150663
10. Liu SH, Shen MY, Kuan CF, Kuan HC, Ke CY, Chiang CL (2019) Improving thermal stability of polyurethane through the addition of hyperbranched polysiloxane. Polymers (Basel) 11(4):1–16
11. Gongalsky MB, Kargina JV, Cruz JF, Sanchez-royo JF, Chirvony VS, Osminkina LA, Sailor MJ (2019) Formation of Si/SiO_2 luminescent quantum dots from mesoporous silicon by sodium tetraborate/citric acid oxidation treatment. Front Chem 7:165
12. Swinehart F (1962) The Beer-Lambert. J Chem Educ 39(7):333–335
13. Khoo HE, Azlan A, Tang ST, Lim SM (2017) Anthocyanidins and anthocyanins: colored pigments as food, pharmaceutical ingredients, and the potential health benefits. Food Nutr Res 61(1):1361779

Technology Comparison of SiO_2 Nanoparticles for Enhanced Oil Recovery in Fang Oilfield

Truong Sinh Le, Ai Chi Vo, and Kreangkrai Maneeintr

1 Introduction

A large portion of the current world oil creation originates from the developed fields. An expanding oil recuperation from the maturing resources is a significant concern about the oil organizations and specialists. Moreover, the rate of substitution of the produced saves by the new revelations has been declining consistently in the most recent decades [1]. The expansion of the recovery factors from the developing fields under the secondary stage like water flooding and the tertiary stage or enhanced oil recovery (EOR) such as chemical EOR (CEOR) will be basic to fulfill the developing industrial need in the coming years [2].

The chemical EOR mechanisms can be a decrease in the interfacial tension and an increase in the mobility ratio [3]. The traditional CEOR strategies have their impediments in that the polymers, whose main recovery mechanism is to increase the viscosity of injections and subsequently the mobility, suffer viscosity loss within the presence of saline solutions and raised temperature conditions. The surfactant and alkali lose their efficiency during their stream in the permeable media because of an adsorption.

Consequently, the various methods of compound flood infusions are conceived, researched, and applied for EOR measures [4]. The nanofluids are made by the dispersion of the nanoparticles to fluids for intensification and improvement of certain properties at low-volume concentrations of the dispersing medium. At that point, the main feature of nanofluids is that their properties greatly depend on the dimensions of nanoparticles of the components [5].

T. S. Le · A. C. Vo · K. Maneeintr (✉)
Carbon Capture, Storage and Utilization Research Group, Department of Mining and Petroleum Engineering, Faculty of Engineering, Chulalongkorn University, Bangkok 10330, Thailand
e-mail: Krengkrai.M@chula.ac.th

R. K. Agarwal (ed.), *Recent Advances in Manufacturing Engineering and Processes*, Lecture Notes in Mechanical Engineering,
https://doi.org/10.1007/978-981-16-3934-0_21

The arrangement of nanoparticles at the interface of oil and water is characterized by their wettability. The wettability of the reservoir rock likewise changes by adsorbing these particles on the rock surfaces. The wettability alteration in the surface of reservoir rock changes the resistances against fluid stream and thus changing the relative permeability of oil and water in the reservoir [6]. Besides, when the nanoparticles are combined with a surfactant, the nanoparticles can be arranged at the interfacial layers and the interfacial tension (IFT) is strongly reduced, resulting in an improvement of the surfactant effectiveness on IFT reduction [7].

Sansai area in Fang oilfield is a small onshore reservoir in Thailand containing oil with gravity ranging from 20 to 40°API and viscosity of approximately 10–120 cP. The total production from Fang basin is approximately 9 million barrels (MMbbl) since early 1960 to the present day [8]. For more than 60 years of primary production, the Fang oilfield has low-oil production. The trouble has been credited to unfavorable production technology. Water flooding has been considered to use in this field. However, the production can be upgraded more with the EOR method. However, the suitable technologies have not been studied for commercial production yet [9]. Therefore, the objectives of this study are to evaluate ability of nanosilica flooding for EOR process in Sansai area, Fang oilfield and compare the oil recovery factor with the conventional water flooding and natural flow by using simulation with ECLIPSE. Moreover, the injection rate and the period of nanoparticle injection are also studied. This study will be used as a reference result to use for the future EOR application in Fang oilfield and the related areas.

2 Simulation

The simulation is run with ECLIPSE software, a product of Schlumberger. This simulator is the most feature-rich and comprehensive reservoir simulator on the market, covering the whole range of the reservoir models, including black oil, compositional, thermal finite-volume and streamline simulation.

2.1 Reservoir Data

The data of Fang oilfield is obtained from the defense energy department (DED), the Northern Petroleum Development Center in Chiang Mai Province. One of the areas in Fang oilfield is Sansai area, and the details of formation and fluid properties in this area are from the previous study [10] and presented in Table 1. Also, Fig. 1 presents the 3D model of porosity distribution of Sansai area from this software. The reservoir model is built in a heterogeneous model with each grid in all reservoir and set with the porosity data ranging from 0.2 to 0.3. An injection well and a production well located at the 2 corners of the reservoir and through from the top layer to the bottom layer for 538.8 m.

Table 1 Reservoir properties using for reservoir model [10]

Parameter	Values	Unit
Grid dimension	25 × 25 × 6	Block
Reservoir size	381 × 381 × 9.144	m
Top of reservoir	1347.216	m
Reservoir thickness	9.144	m
Porosity	0.2–0.3	Fraction
Horizontal permeability	110–190	md
Vertical permeability	11–19	md
Reservoir pressure	6,550	kPa
Reservoir temperature	62.2	°C
Reservoir type	Sandstone	

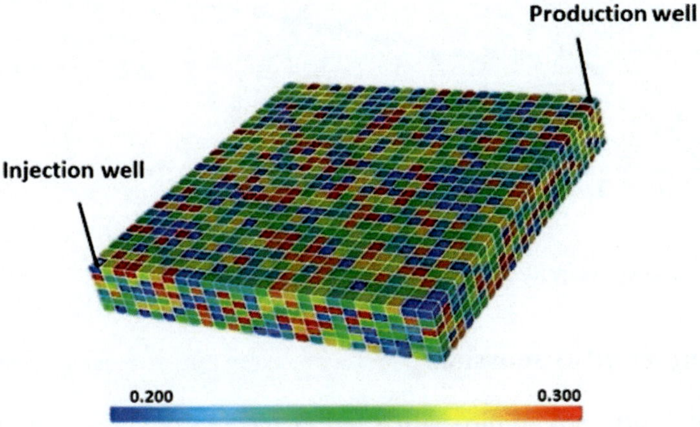

Fig. 1 3D model of porosity distribution in the reservoir

2.2 *Methodology*

This study aims to simulate the 3 main technologies, i.e., natural flow, water flooding, and nanosilica flooding with the injection rate of 79.5 m^3/day and the injection period within 10 years to see the performance and compare the results. Also, the injection rates ranged from 39.75 to 159 m^3/day are evaluated. Besides, the injection period of SiO_2 nanoparticles is another important parameter. It will be injected for 1, 2 and 3 years, followed by water injection until the end of 6 years of injection. More details of study plan are presented in Table 2.

Table 2 Study plan

Parameters	Strategies
Technology	Natural flow, water flooding, nanosilica
Injection rate	39.75–159 m^3/day (250–1,000 barrels/day)
Injection time (years)	1 year nanosilica–5 years water 2 years nanosilica–4 years water 2 years nanosilica–4 years water 3 years nanosilica nanosilica–3 years water

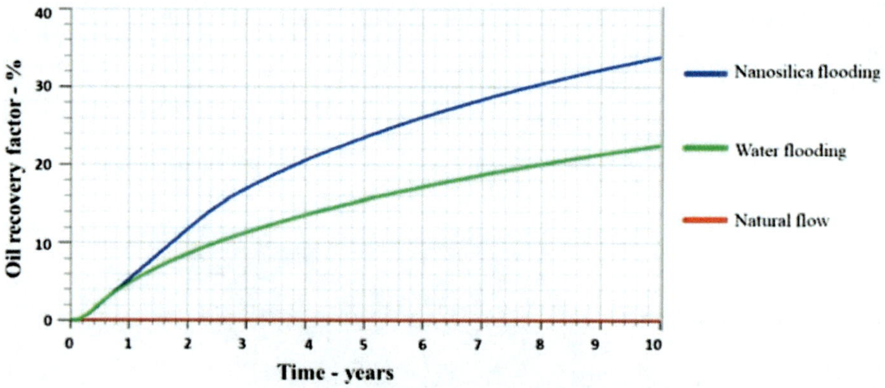

Fig. 2 Technology comparison for oil recovery within 10 years

3 Results and Discussion

3.1 Technology Comparison for Oil Recovery

The reservoir pressure of is 6,650 kPa. It is considered that the pressure is too low to have a natural flow to produce oil from the reservoir. The enhanced oil recovery method should be conducted to get any benefits from this reservoir. As illustrated in Fig. 2, no amount of oil will be produced along 10 years from this natural flow technique. Therefore, the water flooding and nanosilica flooding methods have to be applied. From the results, the recovery factors of water flooding and nanosilica flooding are 22.51% and 33.81%, respectively. These results show that among these 3 techniques, nanosilica can improve the production of oil from the recovery factor for 11.3% higher than that of water flooding.

3.2 Effect of SiO_2 Injection Rate

The injection rate is an important parameter needed to consider because an operating cost will depend on the amount of the fluid injected into the reservoir.

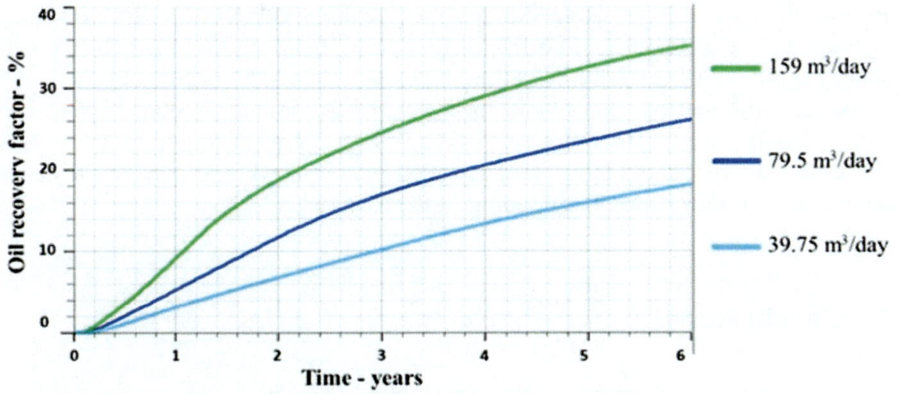

Fig. 3 Result of injection rate of nanosilica

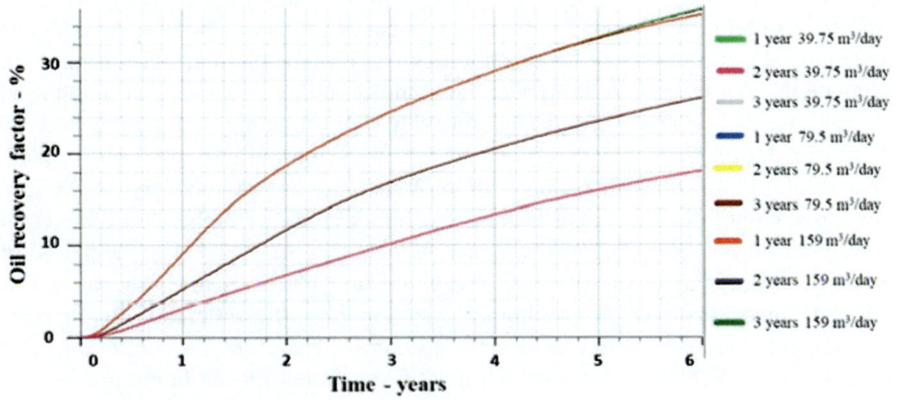

Fig. 4 Result of injection period of nanosilica on oil recovery

Figure 3 shows that the higher injection rate of nanosilica flooding, the higher oil recovery factor can be obtained. The oil recovery factor of 159 m^3/day, 79.5 m^3/day and 39.75 m^3/day are 35.2%, 26.1% and 18.1%, respectively. But the results do not increase with the same ratio with the injection rate. When the injection rate increases from 39.75 to 79.5 m^3/day, the recovery factor increases by 8.0%. Also, when the injection rate increases from 79.5 to 159 m^3/day, the nanosilica fluid can change the recovery factor only 9.1%.

3.3 *Effect of SiO_2 Injection Period*

For the nanosilica injection period from 1, 2 and 3 years of injection and the total of 6 years, the results are presented in Fig. 4. With the injection rate of 39.75 and

79.5 m^3/day, the results at the different period are relatively the same for each injection period at 18.1% and 26.1%, respectively. However, the small change will appear at the high-injection rate. At the injection rate of 159 m^3/day, the 1, 2 and 3 year injection periods can obtain the recovery factor for 35.20%, 35.79% and 35.82%, respectively. The change is less than 1%. Therefore, based on the results and economic viewpoint, the shorter injection period like 1 year may be the better option to start with nanosilica flooding at the lower cost.

4 Conclusion

Over the 60 years of oil production with primary method, Sansai area in Fang oilfield can produce less oil. The production can be improved with the water flooding and enhanced oil recovery (EOR) methods such as chemical EOR (CEOR). However, the technologies that fit well with this area have not been studied for commercial production yet. SiO_2 nanoparticle can be used as a CEOR technique. The utilization of nanotechnology can provide the potential for more oil production. Therefore, this study aims to evaluate ability of nanosilica flooding in Sansai area and compare the production with water flooding and natural flow by using simulation. Also, the injection rate and injection period have been studied. The results show that nanosilica can provide the highest recovery at 33.81% compared to both conventional techniques. In addition, the highest oil production with the injection rates at 159 m^3/day can be obtained for 35.2%. For the period of nanofluid injection, the result of different injection period is insignificant at low injection rate. However, at the higher injection rate, the difference in injection period can be observed with less than 1% difference. Therefore, the shorter injection period can be the better alternative for field development for the future production. Also, this study can be applied for the SiO_2 nanoparticles application in this oilfield in future to enhance oil production.

Acknowledgements The authors would like to thank the scholarship program for ASEAN Countries of Chulalongkorn University and Chevron Thailand Exploration and Production Ltd. for financial support of this project.

References

1. Alvarado V, Manrique E (2010) Enhanced oil recovery: an update review. Energies 3:1529
2. Gbadamosi AO, Kiwalabye J, Junin R, Augustine A (2018) A review of gas enhanced oil recovery schemes use in the North Sea. J Pet Explor Prod Technol 8:1373
3. Thomas S (2008) Enhanced oil recovery-an overview. Oil Gas Sci Technol 63:9
4. Gbadamosi AO, Junin R, Manan MA, Augustine A, Yusuff AS (2019) An overview of chemical enhanced oil recovery: recent advances and prospects. Int Nano Lett 9:171

5. Suleimanov BA, Ismailov FS, Veliyev EF (2011) Nanofluid for enhanced oil recovery. J Pet Sci Eng 78:431
6. Kazemzadeh Y, Shojaei S, Riazi M, Sharifi M (2019) Review on application of nanoparticles for EOR purposes: a critical review of the opportunities and challenges. Chin J Chem Eng 27:237
7. Mohajeri M, Rasaei MR, Hekmatzadeh M (2019) Experimental study on using SiO2 nanoparticles along with surfactant in an EOR process in micromodel. Pet Res 4:59
8. Settakul N (2009) Fang oilfield development. Walailak J Sci Technol 6:1
9. Maneeintr K, Chewaroungroaj J, Boonpramote T (2020) Enhanced oil recovery: potential for oil production development in Fang oilfield, Thailand. IOP conference series: earth environmental science, vol 463, p 012063
10. Maneeintr K, Yoosook H (2018) Simulation of CO2 geological storage together with a water alternating gas process for oil production in fang oilfield (Thailand). Computer aided chemical engineering, vol 43, p 1445

Analysis on Liquid-Crystalline Model and Behavior of Lipid

Yiran Luo

1 Introduction

Generally, liquid crystal (LC) is defined when a substance possesses orientational order and positional disorder simultaneously in its microstructure. Namely, each LC molecule has the same preferred orientation, while being free to move independently in space, resulting in its anisotropy of optical, electrical, and magnetic properties. On a macroscale, the material behaves like a liquid, justified by its fluidity and inability support shear. Meanwhile, it retains certain crystalline features, such as birefringence. In addition, liquid crystal material is sufficiently sensitive to any imposition of electromagnetic field [1]. This peculiar trait allows the LC material to possess favorable steerability and therefore be applied widely in engineering and optics.

The first discovery of liquid crystal was made in the 1880s when Friedrich Reinitzer attempted to measure the melting point of Cholesteryl benzoate. Since then, various sorts of lipid were confirmed of be capable to reach the liquid-crystalline phase. Yet, a common theory to explain lipid's behavior in term of liquid crystal is not achieved, realized by the author, while surveying literature on relevant topic. Also, this topic is of interest and complexity for its interdisciplinarity. As this article primarily discusses the subject from a perspective of physics, this topic can be exploited by biological and chemical means as well. Hence, research result on this subject can also be applied in multiple disciplines. With the necessity of collating and putting together the rather discrete theories and works, studies and works that are related to lipid and liquid crystal will be reviewed from a summative perspective in this paper. With the reference to a variety of literature, including some modern text books, theories mainly adapted from the

Y. Luo (✉)
Shenzhen Senior High School, Shenzhen 518040, Guangdong Province,
People's Republic of China
e-mail: angela@cas-harbour.org

R. K. Agarwal (ed.), *Recent Advances in Manufacturing Engineering and Processes*,
Lecture Notes in Mechanical Engineering,
https://doi.org/10.1007/978-981-16-3934-0_22

works of R. W. Corkery et al. and V. Luzzati et al., and other studies on their rather specialized topic on LC and lipids, this article is proposed to be a brief introduction on the topic of LC and lipids, which is aimed at providing assistance to students and researchers.

2 Basic Theory of Liquid Crystal

Liquid-crystalline phase, also known as the mesophase, is an intermediate phase between solid and liquid. There are two ways to obtain liquid crystal. The first method is to simply melt a crystalline solid. LC produced by this means is called the thermotropic LC. By dissolving a substance in an aqueous environment, another LC named the lyotropic liquid crystal can be obtained. Among these two kinds of methods, LC can be further sorted into different manners, which, as the most common ones, are nematics, cholesterics, and smectics. In a nematic mesophase, all molecules in the substance align toward a distinct orientation, represented by a unit vector, $\overrightarrow{n}(r)$, called the director vector. A cholesteric mesophase have a similar structure. However, the director vector in the cholesteric mesophase varies periodically that it rotates about an axis, to be precise. There is no phase transition between nematics and cholesterics. Ergo, it can be argued that the cholesterics is just a pile of nematics, or that nematics is a type of cholesterics with an infinite pitch. (Pitch is defined as the minimum distance measured between two rotationally in-phase director vectors [2]) Fig. 1 provides an approximate configuration, where the director vectors are drawn from the left profile plane of the illustration.

A smectic mesophase, however, is very different from the former ones. Its most notable feature is that it is separated into layers. Namely, molecules in a smectic mesophase acquire certain positional order, so that individual molecule is unable to

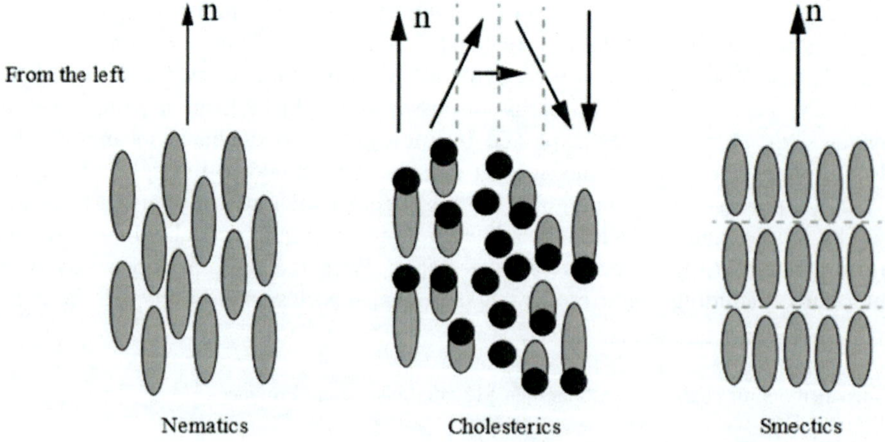

Fig. 1 Illustration of different mesophases

cross between the layers. This feature is manifested by an enormously larger viscosity of the material on a macroscale, which almost forbid the possibility of designing a liquid crystal display (LCD) based on the smectics.

Another classification of mesophase is called columnar mesophase, in which molecules assemble into cylindrical or discotic structure, as shown in Fig. 2.

Elastic property is an equally important property about liquid crystal. There are three types of deformation that can be conducted upon liquid crystal. They are splay, twist, and bend, as shown in Fig. 3. The general formula for computing the elastic energy density caused by the deformation is given as:

$$f_{\text{ela}} = \frac{1}{2}\left[K_{11}(\nabla \cdot \vec{n})^2 + K_{22}(\vec{n} \cdot \nabla \times \vec{n})^2 + K_{33}(\vec{n} \times \nabla \times \vec{n})^2\right] \quad (1)$$

where K_{11}, K_{22}, and K_{33} are Frank elastic constant, $\vec{n}$ is the director vector. These three terms represent the contribution of splay, twist, and bend to the total elastic energy density, respectively [3].

According to study [4], the time required for any liquid crystal to respond to an applied electric field, T, is proportional to the director reorientation time, τ_0, which is given as:

Fig. 2 Illustration of columnar mesophase

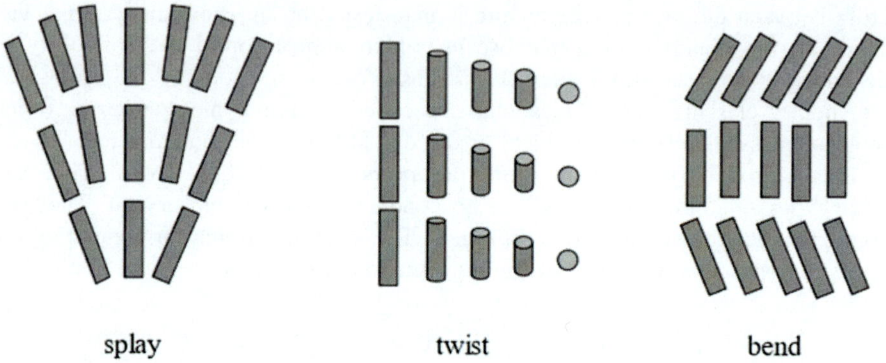

Fig. 3 Illustration of different types of deformation [3]

$$T \propto \tau_0, \ \tau_0 = \frac{\gamma_1 d^2}{K_{33}\pi^2} \tag{2}$$

where γ_1 is the rotational viscosity of the liquid crystal, which is apparently positively correlated with its viscosity. Therefore, a larger viscosity could result in longer delay, and this is an extremely undesired attribute regarding the performance of an LCD. Thus, LCD is barely made from smectics. Nonetheless, this result also suggests that the display can maintain a certain image for a rather long time, even prolonged when the power is off. This is of great prospect concerning the energy conservation issue. Smectic LCD can already be applied to electronic billboard and electronic photo frame, etc. [5].

3 Models of Liquid-Crystalline Lipid in Molten Case

The most readily and intuitional way to obtain LC is through melting crystalline solid. However, lipids are another major source of known LC, due to their unique structural feature: the commonly possessed hydrophobic acyl chains. Here, the author will take one of the most common lipids, triglyceride, as the example for the discussion.

There are some well-accepted postulations, proposed by K. Larsson, for the model of liquid-crystalline, molten triglyceride: First, the triglyceride molecules in the melt form *Y*-shaped conformers, where the 3 fatty acid chains are splayed 120° apart from each other in a plane. Second, the *Y*-shaped conformers are bounded loosely and individually in disk-like volumes. Third, disk-shaped molecules can self-assemble into discotic mesophases. There is evidence that triglyceride molecules can associate with each other when their chains are conformationally and rotationally disordered. The *Y*-shaped configuration is responsible for minimizing the free energy of the system and is justified by Monte Carlo simulation; Similar discotic phase is also observed in analogous thermotropic lipid systems [6], which suggests that the discotic model can be a general model to not only triglyceride, but also a part, if not many, of the multichain lipid molecules. This model can reconcile well the inconsistency between X-ray diffraction observation and previous models.

As the lipid LC, or all LCs, are thermosensitive, research has shown that for lipid, in the vicinity of phase transition, the temperature dependence of the mean orientational order parameter is negatively correlated with the length of the hydrocarbon chains [7]. That the length of the acyl chains is an important parameter is a very natural thought in this model, for the discotic volume is dependent on it.

4 The Liquid-Crystalline Phase of Liquid–Water System

Lipids assemble into different structures as the unit volumes to form liquid-crystalline phases in an aqueous environment due to their amphipathic feature. For a lyotropic LC, concentration is the most dominant parameter. As the concentration varies, lipids form various configurations that can be classified into different phases accordingly.

For the lowest concentration, where the hydrophilic head-groups of the lipids form a circular configuration and the acyl chains point inward (hence the name), a phase called the middle phase is obtained. As the concentration become maximized, the system is altered to a phase named Neat, where the lipids assemble into membrane-like bilayers. In a more general case, the concentration takes value from a range between the maximum and the minimum so that the system takes several intermediate phases in-between Middle and Neat, which is shown in Fig. 4, where the lipids form assorted shape of structures such as rectangle, hexagon, and cube [8].

When temperature is reintroduced as a variable into system at the Neat phase, some variant phases can be acquired. When the temperature is low, the system enters gel phase, where the hydrocarbon chains lose most of their mobility. When the temperature rises, the system behaves like liquid, where the lipids gain

Fig. 4 Illustration of lyotropic LC phases

high-lateral mobility [9]. In between the phases, a smectic ripple phase is observed, which is characterized by corrugations in the lipid bilayer with a well-defined periodicity. The new phase still possesses anisotropy in many aspect [10].

5 Discussion

From a global perspective, studying the behavior of lipid at molten state is studying the phase transition of a particular thermodynamic system. Thermodynamical and statistical methodology can be applied in the theoretical research of the system. In this case, where the molecule is characterized by complex geometry, computer simulation can be of assistance. On the other aspect, the lyotropic LC is often introduced from a biological perspective, typically presented when discussing the membranes of cells and other microorganisms. The study of the type of LC can deepen people's understanding in microbiology. As advancement in biological materials has caught people's attention in recent years, lipid, one of the most commonly used organic products, may become another direction for bio-material exploitation. The theory of lipid's phase transition can also be applied to medical field.

Modern material science also has some effective methodology dealing with this kind of problems, such as the theory of self-assembly. However, most of the mentioned theories for lipid in this paper are phenomenological. Some of them are not mature enough to be quantitative, which is also a consequence of the complexity of the system. Major data, such as the phase-transition temperature, are recorded solely from experiments.

6 Conclusion

In this paper, the basic property and theory of liquid crystal are introduced, and some studies on the liquid-crystalline behavior of lipids are reviewed. Apart from the possession of the general properties of all LC, lipids are of various unique features. The point the author would like to underscore is the significance of acyl chains. In either the thermotropic or the lyotropic case, the mobile chain structure is the source of the liquid-crystalline behavior as they are free to adjust their configuration in space and are able to assemble into assorted structures along with the change in parameter of system. This thesis has collated and reviewed various point of views and theories in order to provide intuition and inspiration for future studies.

Acknowledgements First and foremost, I would like to show my deepest gratitude to my teachers and professors in my school, who have provided me with valuable guidance in every stage of the writing of this thesis. Further, I would like to thank my friends and parents for their encouragement and support. Without all their enlightening instruction and impressive kindness, I could not have completed my thesis.

References

1. Cui X (2014) An all-atom molecular dynamics simulation study of liquid crystal cell. Hebei University of Technology
2. Andrienko D (2018) Introduction to liquid crystals. J Mol Liq 267(4):520–541
3. Yang DK, Wu ST (2014) Fundamentals of liquid crystal devices, 127–48. https://doi.org/10.1002/9781118751992
4. Wang H (2005) Studies of liquid crystal response time. University of Central Florida
5. Kaifa G (2013) The synthesis of colorants for color display of the smectic liquid crystal. East China University of Science and Technology
6. Corkery RW, Rousseau D, Smith P et al (2007) A case for discotic liquid crystals in molten triglycerides. Langmuir 23(13):7241–7246
7. Morrow MR, Whitehead JP, Lu D (1992) Chain-length dependence of lipid bilayer properties near the liquid crystal to gel phase transition. Biophys J 1(63):18–27
8. Luzzati V (1962) The structure of the liquid-crystalline phases of lipid-water systems. J Cell Biol 12(2):207–219
9. Mukherjee S, Maxfield FR (2004) Membrane domains. Annu Rev Cell Dev Biol 20:839–866
10. Kaasgaard T, Leidy C, Crowe JH et al (2003) Temperature-controlled structure and kinetics of ripple phases in one-and two-component supported lipid bilayers. Biophys J 85(1):350–360

Simulation Study of Sodium Dodecyl Benzene Sulfonate as a Surfactant for Enhanced Oil Recovery of Fang Oilfield

Ai Chi Vo, Truong Sinh Le, and Kreangkrai Maneeintr

1 Introduction

Water flooding, a secondary recovery technology to produce oil by injecting water underground and the oil recovery can be increased [1]. Enhanced oil recovery (EOR) is the oil recovery method by injecting the gases or chemicals and/or thermal energy into the reservoir [2]. The EOR process can be classified into thermal and non-thermal recoveries [3]. The mechanism of thermal method is to decrease the viscosity by contributing heat to heavy oil to improve its mobility. The surfactant flooding, one of the non-thermal methods as a part of chemical EOR process, is obtained by adding the chemicals such as sodium dodecyl benzene sulfonate (SDBS) to increase the oil production [4, 5].

Water flooding has become the most widely applied recovery process. In this process, water is injected into the reservoir through an injection well to displace oil in front of it toward the production well [6]. For longer period of production, water flooding becomes uneconomical, and surfactant has been applied in the flooding process to avoid the negative influences of water [7].

The basic theories for surfactant flooding are that this technique creates low-interfacial tension (IFT) to mobilize the residual oil [8]. The liquid surfactant injected into the reservoir is often a complex chemical system, which creates a so-called micelle solution. During the surfactant flooding, it is essential that the complex system forms microemulsions with the residual oil as it supports the decrease in the IFT and increases the mobility [9].

In Thailand, there is an old oilfield called Fang oilfield, a small onshore reservoir. This field has produced oil for over 60 years of natural production, and it has

A. C. Vo · T. S. Le · K. Maneeintr (✉)
Carbon Capture, Storage and Utilization Research Group, Department of Mining and Petroleum Engineering, Faculty of Engineering, Chulalongkorn University, Bangkok 10330, Thailand
e-mail: Krengkrai.M@chula.ac.th

R. K. Agarwal (ed.), *Recent Advances in Manufacturing Engineering and Processes*, Lecture Notes in Mechanical Engineering,
https://doi.org/10.1007/978-981-16-3934-0_23

low-oil recovery with natural production. It is required to have the new techniques to enhance oil production. Therefore, the objectives of this study are to simulate the surfactant flooding by using SDBS as a surfactant with various conditions such as injection rate and period of surfactant injection for Sansai area in Fang oilfield and to compare this technique of oil recovery with the conventional water flooding and natural flow. The results of this study can be contributed for further EOR application in the area as a reference in future.

2 Simulation

2.1 Sansai Area Data

The data of Sansai area in Fang oilfield is obtained from the defense energy department (DED) and the previous study [10]. Sansai area in Fang oilfield has a geological data with the depth from 300 to 1,200 m and the sand thickness from 1 to 7 m. An oil gravity in Fang oilfield is ranged from 20 to 40° API and approximately 10–120 cP for viscosity. The formation data in Sansai area is shown in Table 1.

The 3D pore-scale model of a porous medium is created from the ECLIPSE software program from Schlumberger Company. It is an essential method in quantitative characterization the medium and prediction its transport properties [11]. All grids in reservoir are set at porosity ranged from 0.2 to 0.3 and the permeabilities of the formation are ranged from 110 to 190 mD. The reservoir is built in a heterogeneous model with the injection well and production well located at 2 corners of the reservoir from the top layer to the bottom layer as presented in Fig. 1.

Table 1 Characteristics of Sansai area, Fang oilfield

Parameter	Values	Unit
Grid dimension	25 × 25 × 6	Block
Reservoir size	381 × 381 × 9.144	m
Top of reservoir	1 347.216	m
Reservoir thickness	9.144	m
Porosity	0.2–0.3	Fraction
Horizontal permeability	110–190	md
Vertical permeability	11–19	md
Reservoir pressure	6550	kPa
Reservoir temperature	62.2	°C
Reservoir type	Sandstone	

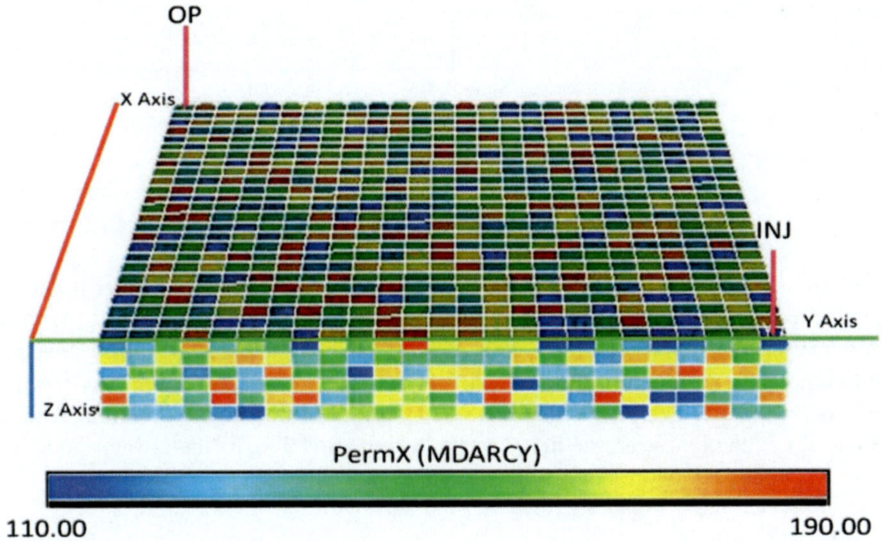

Fig. 1 Permeability distribution in heterogeneous 3D model

2.2 *Methodology*

The use of surfactant technique allows it improving the oilfield performance after water flooding process. The analysis of the chemical injection strategy is necessary for the success of the operation which plays a major role in the efficiency of the recovery process, including the order and the time gap between each chemical slug injection [12].

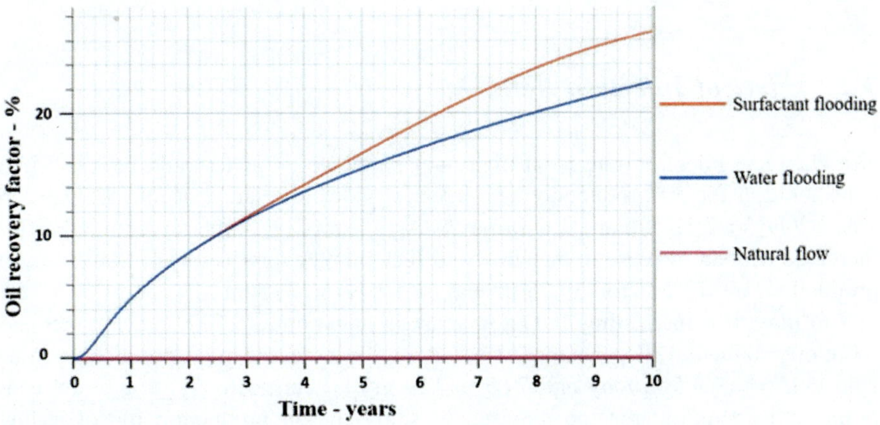

Fig. 2 Technology comparison

Table 2 Injection plan for surfactant flooding in EOR

Parameter	Strategies
Technology	Natural flow, water, surfactant
Injection rate	39.75–159 m^3/day (250–1,000 barrels/day)
Injection period (years)	1 year SDBS–5 years water 2 years SDBS–4 years water 3 years SDBS–3 years water

This study simulates 3 main technologies which are natural flow, water flooding and SDBS surfactant flooding. Also, the effect of the injection rates at 39.75, 79.5 and 159 m^3/day has been studied. In addition, the surfactant will not be injected throughout the production time. The injection period of SDBS from 1 to 3 years will be evaluated. The plan is that the SDBS surfactant flooding will be injected for 3 cases, i.e., the first year, the first 2 years, and the first 3 years then; water flooding will be processed. More details are presented in Table 2.

3 Results and discussion

3.1 Technology comparison

Due to the old field and low pressure in the reservoir at 6,650 kPa, the simulation result from a natural flow technique can produce oil from the reservoir at comparatively 0 % for the period of 10 years. It can be seen clearly that water flooding and surfactant flooding should be applied to increase more oil production. As shown in Fig. 2, with the injection rate of 79.5 m3/day, the recovery factors are 26.7 % for SDBS surfactant flooding and 22.5 % for water flooding. As a result, among these techniques, the SDBS surfactant can be used for EOR with recovery factor of 18.7% higher in comparison with water flooding.

3.2 Effect of Injection Rate

The chemical injection rate is an important parameter which should be taken into consideration for oil production. The injection rates are varied at 39.75, 79.5 and 159 m^3/day or 250, 500 and 1,000 barrels/day. From Fig. 3, the oil recovery factor increases with an increase in injection rate. The oil recovery factors of SDBS for oil production are 12.2% for 39.75 m^3/day, 19.2% for 79.5 m^3/day and 31.6% for 159 m^3/day. It is interesting that an increase in injection rate 100% cannot increase more oil production. For instance, 7% and 12.4% increases in oil production come from 39.75 to 79.5 m^3/day and 79.5 to 159 m^3/day, respectively. Therefore, economic consideration may be required to study for oil production by increasing injection rate.

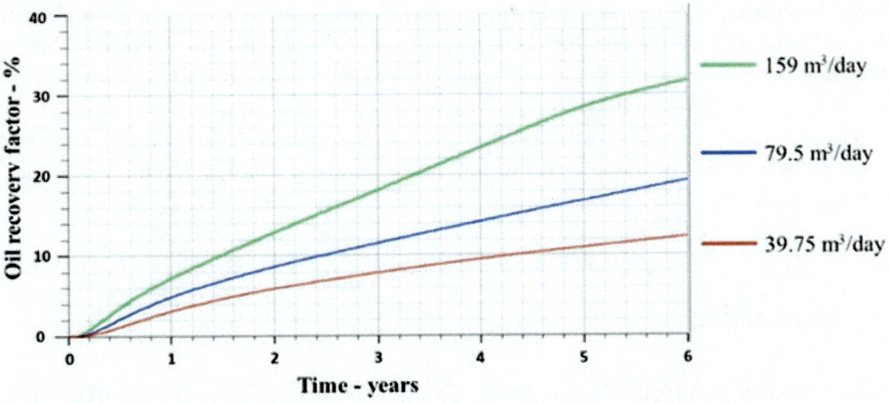

Fig. 3 Result of SDBS injection rate

3.3 *Effect of Injection Period*

An injection period is the plan for the period of SDBS injection to enhance oil production. The total period of injection is 6 years by injection of SDBS for the first injection, followed by injection of water. There are 3 plans of this injection. Plan A is 1 year of SDBS injection and 5 years of water injection. Plan B is 2 years of SDBS injection and 4 years of water injection, and Plan C is 3 years of SDBS injection and 3 years of water injection. The injection rate for this study is 159 m^3/day because it provides the highest oil production.

Figure 4 shows the simulation results of the effect of injection period. It is presented that for the first 3 years of injection, and the oil production is relatively the same. However, after 3.5 year Plan A with 1 year starts to deviate from the group and provide the lower oil recovery. Later, after 5 years, Plan B and C will

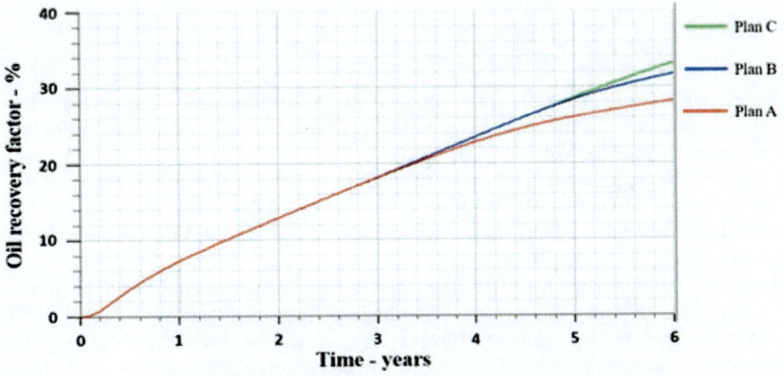

Fig. 4 Result of injection period on oil production

have the difference in oil production. After 6 years of production, Plan C with 3 years injection provides the better result in term of oil production. The oil recovery factors are 28.1%, 31.6% and 33.1% for Plan A, B and C, respectively. However, the difference in oil recovery factor is less than 5% for all cases. Therefore, it can be recommended that Plan A might be better strategy in term of the economic point of view for oil production because it can provide the lowest cost among them.

4 Conclusion

Sansai area in Fang oilfield has produced oil with natural flow for over 60 years. Water flooding and chemical enhanced oil recovery technologies can be applied to increase more oil production. This work studies the water flooding and surfactant injection with sodium dodecyl benzene sulfonate (SDBS) by using the simulation with ECLIPSE. Also, the injection rate and period of injection for Sansai area in Fang oilfield have been studied and compared. The simulation results show that among 3 technologies, the surfactant flooding by using SDBS provides the highest recovery factor at 26.7%. Also, an oil recovery can increase with an increase in injection rate. The injection rate at 159 m^3/day shows the highest recovery factor at 31.6%. For the injection period of 1, 2 and 3 years of SDBS injection, the results illustrate that 3-year injection can offer the highest oil recovery at 33.1%. Nevertheless, the difference in the oil recovery factor is less than 5%. Consequently, the optimum condition can be 1-year injection period to save more cost. This study can be considered for the further study in this field in future.

Acknowledgements The authors would like to thank the scholarship program for ASEAN Countries of Chulalongkorn University and Chevron Thailand Exploration and Production Ltd. for financial support of this project.

References

1. Maneeintr K, Chewaroungroaj J, Boonpramote T (2020) Enhanced oil recovery: potential for oil production development in Fang oilfield, Thailand. IOP conference series: earth and environmental science, vol 463, p 012063
2. Thomas S (2007) Enhanced oil recovery—an overview. Oil Gas Sci Technol 63:9
3. Esmail MAM, Mohamed H, Zubairu A, Raghib S, Habib MA, Mohamed M (2019) A comprehensive review of thermal enhanced oil recovery: techniques evaluation. J Energy Resour Technol 141:030801–1
4. Chuaicham C, Maneeintr K (2017) Interfacial tension measurement of light oil from thailand to enhance oil recovery for surfactant flooding application. Mater Sci Forum 890:235
5. Zabihi S, Faraji D, Rahnama Y, Hezave AZ, Ayatollahi S (2020) Relative permeability measurement in carbonate rocks, the effects of conventional surfactants vs. Ionic liquid-based surfactants. J Dispersion Sci Technol 41:1797

6. Arab D, Kantzas A, Bryant SL (2020) Water flooding of oil reservoirs: Effect of oil viscosity and injection velocity on the interplay between capillary and viscous forces. J Pet Sci Eng 186:106691
7. Zhou X, Wang Y, Zhang L, Zhang K, Jiang Q, Pu H, Wang L, Yuan Q (2020) Evaluation of enhanced oil recovery potential using gas/water flooding in a tight oil reservoir. Fuel 272:117706
8. Druetta P, Picchioni F (2020) Surfactant flooding: the influence of the physical properties on the recovery efficiency. Petroleum 6:149
9. Sandersen SB (2012) Enhanced oil recovery with surfactant flooding. Ph.D. thesis, Technical University of Denmark, Denmark
10. Maneeintr K, Yoosook H (2018) Simulation of CO2 geological storage together with a water alternating gas process for oil production in Fang oilfield (Thailand). Computer aided chemical engineering, vol 43, p 1445
11. Jiang Z, van Dijke MIJ, Sorbie KS, Couples GD (2013) Representation of multiscale heterogeneity via multiscale pore networks. Water Resour Res 49:5437
12. Druetta P, Picchioni F (2018) Surfactant–polymer flooding: influence of the injection scheme. Energy Fuels 32:12231